The Charisma Machine

Infrastructures Series

Edited by Geoffrey C. Bowker and Paul N. Edwards

The Charisma Machine

The Life, Death, and Legacy of One Laptop per Child

Morgan G. Ames

The MIT Press
Cambridge, Massachusetts
London, England

This book was set in ITC Stone Serif Std and ITC Stone Sans Std by Toppan Best-set Premedia Limited. Printed and bound in the United States of America.

Library of Congress Cataloging-in-Publication Data

Names: Ames, Morgan G., author.
Title: The charisma machine : the life, death, and legacy of One Laptop Per Child / Morgan G. Ames.
Description: Cambridge, MA : MIT Press, [2019] | Series: Infrastructures | Includes bibliographical references and index.
Identifiers: LCCN 2018050938 | ISBN 9780262537445 (pbk. : alk. paper)
Subjects: LCSH: Laptop computers--Developing countries. | Computers and children--Developing countries.
Classification: LCC QA76.5 .A4255 2019 | DDC 004.1609724--dc23 LC record available at https://lccn.loc.gov/2018050938

10 9 8 7 6 5 4 3 2 1

For Carla

Contents

Acknowledgments

This project has made powerfully concrete to me that intellectual development is a profoundly social process. Just as the motivations of the Paraguayan children whom I observed were influenced by their teachers, parents, and friends, my own research trajectory has been deeply influenced by many important people in my life. Indeed, at times it seems strange that research projects have ownership when we all owe so much to our intellectual communities.

First and foremost, I thank my graduate advisor, Fred Turner, for his intellectual guidance. Though our topics, methods, and even writing habits may differ, I have learned so much from him about how to navigate the halls of academe and to produce rigorous, intellectually honest, and (hopefully) insightful research—and work that reflects my own passions. Words cannot express the gratitude I feel for this intellectual and social guidance. My committee has also helped me along this path: Cliff Nass, Jeremy Bailenson, Tanya Luhrmann, and John Willinsky. I especially honor the memory of Cliff, who in many ways was like a second advisor. He hooded me in place of Fred in June 2013 and was always available to provide ever-upbeat and ever-insightful feedback. We miss you, Cliff.

Just as important to this research are my colleagues and participants in Paraguay. My fieldwork there was one of the most challenging and enlightening experiences of my life. The gracious hospitality and support of the employees of Paraguay Educa, especially Pacita, enabled my fieldwork. I also owe so much to my research assistant, Liliana, who not only came along to many of the house visits and interviews to help me with my nonexistent Guaraní but scheduled and transcribed many of my interviews, and moreover welcomed me into her home as if I were a member of her family.

Her superb local knowledge and many friends made my fieldwork so much easier. I want to thank all of the teachers, students, parents, and others for allowing me to shadow their experiences and for patiently answering my many questions. Geeking out with Bernie, Sebastián, Martín, and the technical team gave me a welcome respite during my time in Paraguay. Bernie also provided a welcome counterpoint to my observations and theories in Paraguay and beyond. I honor the memory of Carla, who was not only my host but my friend. When cancer took her from us much, much too early, it took one of the brightest and most passionate advocates for Paraguayan children.

The lion's share of day-to-day support came from my cowriters over the years: ShinJoung Yeo, Daniela Rosner, Lilly Irani, Megan Finn, Lilly Nguyen, Tricia Wang, Lauren Schmidt, Christine Larson, Anita Varma, John Alaníz, Mark Gardiner, and more. These incredible scholars joined me to write in cafés and libraries all over the San Francisco Bay Area, and most also offered generous feedback on chapter drafts along the way. I have come to learn that I love to write socially, and this group has seen me through. Equally important was the generous community of scholars who offered feedback on chapters throughout the writing process. The dissertation on which this book is based was honed with the help of the Stanford Humanities Center dissertation writing group, convened by Katja Zelljadt. The two years I spent in this talented, interdisciplinary group gave me countless ideas for how to both broaden my work to appeal to more audiences and deepen it to make it more theoretically impactful. More recently, I have bent the ear of insightful colleagues including Joseph Klett, Matt Rafalow, Amber Levinson, Alicia Blum-Ross, Antero Garcia, Jenna Burrell, Damien Droney, Anne Jonas, Richmond Wong, Noura Howell, Christoph Derndorfer, and Nick Merrill, who all provided constructive feedback on (in many cases) multiple chapters. Finally, Laura Portwood-Stacer of Manuscript Works helped me reframe an early draft, Georgia Saltsman and Charlotte Nix helped me proof the near-final draft, and anonymous reviewers gave me invaluable feedback in between, guided by the encouraging editorial leadership of Katie Helke at the MIT Press.

This project was shaped by a number of people in the broader academic community. I thank my first postdoctoral supervisor, Paul Dourish, and the engaging community at the University of California, Irvine to which he introduced me, especially Chris Wolf, Noopur Raval, Silvia Lindtner, Katie

Pine, Katherine Lo, Melissa Gregg, and the other members of the Intel Science and Technology Center for Social Computing. I hope the technology industry will again support such generative collaborations as the ISTC produced. I thank Eden Medina for including me in an incredible group of scholars studying science and technology in Latin America, and Juan Rodriguez and Hector Danilo Fernandez L'Hoeste for helping me sustain these connections. I have also benefited from the leaders and participants of the Values in Design workshops in 2008 and 2010, the Consortium for the Science of Sociotechnical Systems in 2011, the Digital Media and Learning Summer Fellows program in 2012, and the iConference Doctoral Colloquium in 2013.

I have saved the most important people in my life for last. To Josh, my wonderful partner: your faith in me has truly sustained me, and your commitment to an equal partnership has made this intellectual labor possible. Our relationship started around the time I embarked on this project, and I am so grateful for the strength you have helped me develop through the last decade to see it through. And to Mom and Dad, I credit you with nurturing my intellectual curiosity and cultivating a joy for living and learning throughout my youth. Your unconditional love and unwavering support mean more to me than I could ever express.

Introduction

On January 26, 2005, capitalists and leaders from around the world made their annual pilgrimage to Davos, Switzerland, for the World Economic Forum. A decade before, those worshipping at this altar of neoliberal globalization had supported missions to spread computers and connectivity to the far corners of the world, such as the Technology to Alleviate Poverty project and the Digital Divide Initiative, but the dot-com crash in 2001 had largely quashed the appeal of those kinds of technological visions.

Even so, Nicholas Negroponte, a professor at the Massachusetts Institute of Technology (MIT), tested the waters for rekindling that vision under a new name. He plied the halls of Davos that January with a crude mock-up and what he hoped would be a compelling story of a hundred-dollar laptop for children across the Global South.[1] It would be cheap, it would be powerful, and it would be rolled out by the hundreds of millions to entire countries. He boasted that the project already had a backer—Advanced Micro Devices, a computer processor manufacturer—and interest from other technology companies. He had a vision, and he had the support of a number of fellow MIT professors—he just needed a market.

Negroponte was no stranger to sweeping digital dreams. In 1985, he had cofounded MIT's Media Lab, which had a mission to "invent the future" as well as a penchant for flashy demonstrations of its "big ideas" for corporate sponsors, whose donations gave them access to all of the lab's findings.[2] In 1992, he became a founding investor in *Wired* magazine, for which he wrote a column on "Bits and Atoms"—and reasons to transcend the boundary between them—from 1993 to 1998. These columns informed Negroponte's 1995 book *Being Digital*, which detailed his vision for on-demand digital consumer content, personalized newsfeeds, and what would later be called the "internet of things." The relentless utopianism of *Being Digital* captured the excitement that many felt about the "electronic frontier," the burgeoning online world of the 1990s.[3] It became a best seller and was translated into some forty languages. Negroponte largely shrugged off the sharp criticism his book and columns drew from some scholars, such as legal scholar Cass Sunstein, who has decried the echo chambers of Negroponte's customized "Daily Me" newsfeed idea for polarizing the US political landscape, and cultural historian Fred Turner, who has linked Negroponte's digital boosterism to the commodification of "New Communalist" utopianism in the 1970s and beyond.[4]

A decade later, in January 2005, Negroponte seemed to receive a relatively cool reception in Davos. His hallway pitch for a hundred-dollar laptop garnered a brief blog mention by Travis Kalanick, who was attending the World Economic Forum as a "technology pioneer" (and would later start the ride-sharing platform Uber), but this mention was more due to Negroponte's other accomplishments than his idea for a hundred-dollar laptop. *New York Times* technology journalist John Markoff took up Negroponte's pitch in more depth but concluded that Negroponte had not been given more of an official platform because the forum had moved on from the ideal of closing the digital divide to solving more "fundamental" inequalities.[5]

This tone changed considerably the following November, when Negroponte took the stage at the World Summit on the Information Society in Tunis. Joining United Nations secretary-general Kofi Annan in a presentation in the Access2Democracy session on November 16, Negroponte unveiled an updated mock-up, which he called the "green machine": a bright green plastic laptop about the size of a hardback book. It was outlined in black rubber, sported a bright yellow hand crank that extended from the hinge between keyboard and screen, and had a printed picture of a group of grinning, brown-skinned children in place of a screen.[6] Just one minute of turning the hand crank, Negroponte explained, would give the ultra-low-power laptop forty minutes of charge, making it usable even where there were no power sources.[7]

Despite its low power consumption, the device would still be a fully featured computer, ruggedized to withstand children's explorations—and it would cost a mere one hundred dollars. The laptop was only available to country governments, and Negroponte claimed that many had already "expressed serious interest," even with a high cost of entry: the purchase of at least one million machines.[8] He wanted hundreds of millions to be in use by the end of 2007 and for every child in the Global South to have one by 2010.[9] During the presentation, Annan demonstrated the hand crank on the mock-up—and promptly broke it off.[10]

This presentation, its technical difficulties notwithstanding, lit up journalists' wires and captured imaginations around the world.[11] Even articles skeptical of the project's feasibility generally presented its vision as compelling.[12] The most vocal critics were the African delegation to Tunisia, which as a group was largely unimpressed by the demonstration and voiced concerns about project and infrastructure costs, resources diverted from clean water and better schools, and machine obsolescence and recycling.[13] But these voices were quickly marginalized within the broader discourse around the hundred-dollar laptop. This paved the way for Negroponte, who was on leave from MIT to work full time on this project, to give a marathon of similar presentations over the next two years. On January 27, 2006, Negroponte received a much more enthusiastic welcome at Davos than a year prior. At his presentation, he and Kemal Derviş, an administrator at the United Nations Development Programme, signed an agreement that the UNDP would help distribute the machines.[14] In the months following, Negroponte gave a talk to technology-world luminaries at the Technology, Entertainment, Design (TED) conference, delivered a keynote talk at the World Congress on Information Technology, and spoke at many more conferences and meetings, spinning visions of technologically driven transformation while repeating the mantra "it's an education project, not a laptop project."[15]

Buoyed by these presentations, coverage of what came to be known as One Laptop per Child (OLPC) continued to be largely positive, even as the project struggled to fulfill Negroponte's promises. Two of the features that had generated the most excitement at the November 2005 demonstration in Tunis were quietly rolled back. First, there would be no hand crank; these computers would ship with an AC adapter, like any other device, collapsing the vision of the laptop leapfrogging past regional infrastructural deficiencies. Second, the laptop's cost would be nearly two hundred dollars, not one hundred—and that did not account for costs of infrastructure (such as power and internet access) and maintenance. Moreover, Negroponte's claim of "serious interest" from countries around the world did not result in any orders beyond a few trial machines throughout 2006 and much of 2007. Sales never did reach the level that Negroponte had promised. To date, OLPC has sold less than three million laptops, more than 80% of them to projects in Latin America—far from the hundreds of millions that Negroponte envisioned.

And yet, these realities seemed not to matter. After Negroponte and Annan's presentation in November 2005, OLPC and its charismatic laptop, the XO, quickly became the darlings of the technology world. The project's 2005 funders, Advanced Micro Devices, Google, and News Corporation, were followed in 2006 by eBay, Marvell Technology Group, SES Astra, and Nortel, each of which reportedly contributed two million dollars to the initiative.[16] Some companies also contributed employee time and expertise, such as Red Hat, Google, Brightstar, and Taiwanese companies Quanta (which eventually produced the laptop) and Chi Mei (which manufactured its display).[17] Members of the open source software community, wooed by the promise of a generation of children raised on free software, enthusiastically contributed to the development of OLPC's custom-built software platform, Sugar.

In place of the abandoned hand crank, Negroponte filled his presentation with unsubstantiated stories of children teaching themselves English and their parents to read, of impromptu laptop-enabled classrooms under trees, and of laptop screens being the only light source in a village (never mind how the laptops were recharged). These claims echoed enthusiastically and largely uncritically across the media and technology worlds.[18] Some are still repeated today, even in the face of ample evidence that the project has not lived up to its promises—just as stories about the laptop's hand crank continued to circulate long after the crank idea had been scrapped. These claims became part of the allure of One Laptop per Child's machine. As with many who lead development projects, Negroponte and OLPC's other leaders and contributors wanted to transform the world—not only for what they believed would be for the better but, as we will see, in their own image. "Technology is the only means to educate children in the developing world," Negroponte told the *MIT Technology Review* in October 2005. Later, he said that the laptop project "is probably the only hope. I don't want to place too much on OLPC, but if I really had to look at how to eliminate poverty, create peace, and work on the environment, I can't think of a better way to do it."[19]

...

One Laptop per Child serves as a case study in the complicated consequences of technological utopianism. The puzzle of this book is to untangle what made this project and its laptop so captivating and even the most outrageous claims about it so compelling. Despite OLPC's high profile, hailing from the MIT Media Lab and becoming known around the world, very little is known about how the project's laptops have been used day to day and what impact they have had—a gap this book will fill. But the reasons this is important go beyond mere historical interest. We will see that the same utopian impulses that inspired OLPC had also inspired previous starry-eyed projects in education and development—and they have continued to inspire subsequent projects, from massive open online courses (or MOOCs) to makerspaces, from technology-centric charter schools to coding boot camps. All of these projects have had material effects in the world, some positive, some negative. Although I will explore these dimensions, my primary goal here is not to pass judgment on what counts as a good intervention. More importantly, all of them have failed to achieve their utopian goals—yet it seems as if many technologists, designers, educators, policy makers, and others have failed to learn lessons from them, instead remaining moonstruck about the potential for technological transformation.

This book is thus more than just an account of One Laptop per Child. It is a cautionary tale about technology hype that explains how technologies become charismatic and what the consequences of that charisma can be. We will reach a half century into the past and across the globe to critically examine the consequences of utopia-inspired design, technology's role in play and learning, and the sometimes-fuzzy divide between education and entertainment. We will begin with this central question: why did so many so enthusiastically accept Negroponte's and OLPC's claims—especially when similar promises had been made and broken before? Then, we will explore how these promises were kept, broken, or transformed when OLPC's laptops were put to use. Were the charismatic visions of OLPC compelling—or even recognizable—to the project's intended audience of children in the Global South? Finally, we will examine OLPC's legacy. How have the same promises lived on in new projects, even after the dissolution of the original One Laptop per Child foundation and its apparent failure to achieve its lofty goals?

The Case for Charismatic Technology

Social theory can help us understand, and keep in perspective, the holding power that One Laptop per Child and its laptop have had on technologists and others around the world. Part of OLPC's allure was Negroponte's stories about what the project would accomplish in the world. These referenced some powerful cultural *mythologies*, or foundational narratives that are ritualistically circulated within groups to reinforce collective beliefs. Mythologies have an element of enchantment to them, making certain futures appear at once magical and inevitable, straightforward and divine.[20] Vincent Mosco and David Nye have each shown that for nearly two hundred years, mythologies have been central to the way that the United States and Europe think about technology, connecting what might otherwise be a rather mundane artifact to feelings of awe, transcendence, and a sense of greater purpose—or what they call the *technological sublime*.[21] Paul Dourish and Genevieve Bell juxtapose mythologies against the "messiness" and contested nature of everyday life and conclude, like Mosco, that mythologies are much more than simple falsehoods; they "animate individuals and societies by providing paths to transcendence."[22]

These mythologies invoke certain *social imaginaries* that define our collective social existence, and social imaginaries of childhood and technology's role in it are especially important to OLPC's story. Social imaginaries were originally theorized by Benedict Anderson to explain the "imagined political community" that creates feelings of nationalism—the set of myths, stories, and group identities that, for instance, make "being American" distinct from "being Canadian," despite intertwined histories and similar material conditions of life in the two countries.[23] This book builds on later scholarship that understands a social imaginary more broadly as a set of coherent visions by a group of people to collectively "imagine their social existence," as philosopher Charles Taylor puts it—the ways that people imagine themselves as part of a group and the identities that this group takes on in their minds.[24] Central to a social imaginary is the role of the imagination, as well as the requirement that imaginings be collective.[25] Imaginaries go beyond individual and fleeting fancies; they are coherent and often powerful ideas held by groups of people (whether members of a nation or some other social world, such as hackers or open-source software proponents). Imaginaries also have a normative side that shows us, as Sheila Jasanoff and Sang-Hyun Kim put it, "how life ought, or ought not, to be lived."[26] Moreover, they direct our actions—on an individual level but especially on a group level (e.g., through national policy or group norms)—toward such ends.[27] In the first chapter, we will explore how OLPC's promises were linked to broader imaginaries of childhood, creativity, play, and learning, and in chapter 2, we will see how these imaginaries influenced the design of OLPC's laptop. Whether these imaginaries can be "translated" to children in the Global South will be a guiding question in chapters 3 through 6.

This account also takes seriously the role of media in creating and sustaining these social imaginaries. More than mere pastime, media help shape our shared understandings of the world. As Arjun Appadurai explains, people "see their lives through the prisms of the possible lives offered by mass media in all their forms."[28] OLPC invoked particular social imaginaries in part through a strong media presence: the high-profile talks by Negroponte and other OLPC leaders spawned many news articles and other commentary that largely adopted the project's worldview and echoed its claims, as we will see in chapters 1 and 2. The media will likewise be a potent force among the child beneficiaries of the project, though not in the way that OLPC imagined—one that, instead, may well lead to what Lauren Berlant calls "cruel optimism," wherein unachievable fantasies and desires can become obstacles to living in the present, as chapters 4 and especially 5 will highlight.[29]

These cultural mythologies and social imaginaries are aspects of what social theorists call *ideologies*: the frameworks of norms, generally taken for granted and unconsciously held, that shape our beliefs and practices and that justify differences in power between various social groups.[30] Cultural theorist Stuart Hall describes an ideology as a "system for coding reality" or "a way of representing the order of things which endow[s] its limited perspectives with that natural or divine inevitability which makes them appear universal, natural and coterminous with 'reality' itself."[31] What is important in this definition is the way that ideology fades into the background: by one metaphor commonly used in anthropology, ideologies are as invisible to many people as we imagine water is to a fish.[32]

Some branches of social theory tie ideologies to the operation of state power, just as Benedict Anderson connected social imaginaries to feelings of belonging to a state. The meaning of "ideology" that this book uses is not specific to statehood but is no less powerful a force in people's lives. We live with many ideologies, reinforced across our sociopolitical landscape: neoliberal economics, patriarchal social structures, and Judeo-Christian ethics are among the dominant ideologies in the United States, for instance. In Hall's words, each of these ideologies "works" because it "represses any recognition of the contingency of the historical conditions on which all social relations depend. It represents them, instead, as outside of history: unchangeable, inevitable and natural."[33]

Many of those speaking on behalf of OLPC were certainly discussing the project's world-changing goals as "inevitable and natural": as long as the project managed to get its laptops out in the world, success would follow. This book interrogates that claim. It first analyzes the social imaginaries that influenced the design of OLPC's laptop and how those social imaginaries were meant to be encoded in the laptop itself. It then follows the laptop out into the world to see whether those social imaginaries really were able to travel with it—and what other imaginaries factored into how the laptop was understood in use. This endeavor requires a social theory that can move with the laptop itself—something that can account for how these ideological commitments were designed and built into the laptop and how they traveled with it, while also accounting for the organizational, infrastructural, and cultural narratives that it either invoked or contradicted.

For this, I borrow an idea that originated with the study of populist and religious movements: *charisma*. Max Weber, one of the founders of modern sociology, used charisma to explain the exceptional, even magical, authority that religious leaders seem to have over followers—an authority he witnessed firsthand in the German Empire during the lead-up to World War I. This *charismatic authority* stood in contrast to the other types of authority that Weber was interested in understanding, in particular traditional authority and legal/rational authority. Charisma is not legitimized through bureaucratic or rational means but by followers' belief that a leader has extraordinary, even divine, powers that are not available to ordinary people.[34]

Although a few social theorists have invoked "charisma" to describe technologies or ideas, charisma is generally used to describe the power of humans, especially religious and cult figures, not objects such as laptops.[35] (Nicholas Negroponte, for instance, is a leader that some have found very charismatic.) However, the field of science and technology studies (STS) has shown us that sociological concepts that have been useful in understanding humans can also be used to explain the "actions" of nonhuman actors, such as scientific instruments, technological objects, or even scientific theories.[36] Indeed, these actors can and often do take on lives of their own beyond the intentions of their creators.[37] As Bruno Latour's humorous analysis of the complicated dance between an automatic "door-closer" and the humans interacting with it shows, these nonhuman actors have *agency*—they can act on the world beyond the intentions of their designers—and the way that they reflect various ideological commitments can and often does shift between design and use.[38] By this token, analyzing OLPC's laptops as active *subjects*, not just passive *objects*, becomes a necessary component of the work we will undertake here.

Treating an object as a charismatic subject with some degree of agency borrows perspectives from actor–network theory (ANT), a theory in STS that uses the same analytical lens to understand both human and nonhuman *actors* within a mutually constituted *network* of relations. All actors, human and nonhuman alike, are products of a number of social choices and technical constraints and capabilities (a combination that is often referred to as "sociotechnical" in STS) and in turn shape future sociotechnical relations, a process that STS scholars call "coproduction."[39] This book brings actor–network theory, which provides tools for analyzing the "scripts" that designers build into technologies but tends to favor materialism and actor-centered accounts of the world, into conversation with cultural studies, which focuses more on the beliefs that underlie these networks.[40] In this way, charisma theorizes a specifically utopian "circuit of culture" and provides the means to account for how ideological frameworks inhabit and animate these tools.[41]

In this account, a *charismatic technology* derives its power experientially and symbolically through the possibility or promise of action: what is important is not what the object is but how it invokes the imagination through what it promises to do.[42] The material form of a charismatic technology may be part of this but is less important than a technology's ideological commitments—its "charismatic promises." This means that a charismatic technology does not even need to be present or possessed to have effects. A charismatic technology is thus an active subject within a sociotechnical web of other actors and social imaginaries—and its promises can be enthralling. As sociologist Donald McIntosh explains, "charisma is not so much a quality as an experience. The charismatic object or person is experienced as possessed by and transmitting an uncanny and compelling force."[43]

Charisma moreover implies at least some degree of *persistence* of this compelling force even when an object's form or actions do not match its promises. This is part of the magic of charisma and where a charismatic technology's link to religious experience is especially strong: charisma taps into an ideological framework that is at least partially maintained and strengthened outside (or even counter to) rational thought or evidence. This is also one of the places where the seams between charisma and the world are most visible, as a charismatic technology's acolytes maintain their devotion even in the face of contradictions.[44]

In their promises of action, however, charismatic technologies are deceptive: they make both technological adoption and social change appear straightforward instead of as a difficult process fraught with choices and politics. This gives charismatic technologies a spirit of *technological determinism* (or "technological solutionism"), whereby progress that a technology is supposed to cause is framed as natural and inevitable, thus overriding individual, social, institutional, or other kinds of agency—much like the "exceptionalism" of Weber's charismatic leaders.[45] Scholars in STS have been studying technological determinism for decades now and have sussed out the many ways it is wrong.[46] But outside of those circles—and particularly in the technology development and design world—the belief is still commonplace. It can lead believers to underestimate the sustained commitments (social, political, financial, infrastructural, etc.) needed for both technological adoption and social change.

By the same token, charisma's naturalizing force can make critique and debate appear unnatural—they are, after all, up against what appears to be a natural and inevitable path toward technologically determined progress. Historians have shown us, for instance, that when the United States was building its rail network in the mid-nineteenth century, the feelings of sublime awe and transcendence that the locomotive evoked across the nation led the United States to pay an enormous price in resources and lives in an attempt to realize the utopian promises of rail: the end of wars and parochialism, the merging of cultures, the very "annihilation of space and time."[47] Mosco and others have traced the sublime paths of a host of other technologies: radio, film, atomic energy—and of course computers and the internet.[48] Although charisma is distinct from the technological sublime—it does not necessarily subsume reason with feelings of overwhelming awe—it is similar in upholding a belief that a technology such as OLPC's laptop and its specially-designed software can provide a shortcut to peace and prosperity, even if governments do not actively recognize its potential for this. This also reflects a technologically determinist faith in the power of the laptop itself to create change. Even though over four decades' worth of scholarship on technological determinism has explored in detail its many pitfalls, it is still prevalent in popular culture and especially in technological circles.[49]

However, charisma contains an irony, and it is here that the conception of charismatic technology that I have laid out departs from Weber's account. A charismatic technology may promise to transform its users' sociotechnical existence for the better, but this book will show that charisma is, at heart, ideologically conservative. Charismatic leaders confirm and amplify their audiences' existing ideologies to cultivate their appeal, even as they may paint visions of a better world.[50] A charismatic technology's appeal likewise confirms the value of existing stereotypes, institutions, and power relations. This unchallenging familiarity is what makes a charismatic technology alluring: even as it promises certain benefits, it simultaneously confirms that the ideological worldview of its audience is already right—the charismatic technology will simply amplify it. By referencing and reinforcing existing ideological norms, charismatic technologies can also, by extension, appear unchangeable, inevitable, and natural—and their promises may still hold power even when a technology does not deliver.

One question, of course, is what happens when a charismatic technology comes in contact with a group that does not share the same ideological frame (we will explore this in depth in the later chapters of this book). Another question is what happens when charisma fades. In Weber's account, the charisma of leaders is ultimately unstable. Weber details a process that he calls "routinization," wherein charismatic authority is supplanted by another form of authority such as traditional hierarchies, legal frameworks, or bureaucratic procedures. What happens when the charisma of a particular technology or technological project fades? In this case, OLPC's charismatic authority was not routinized in the classic Weberian sense: the project became nostalgized among its biggest boosters as its authority largely faded, as it retreated from messy realities back to myth. However, the ideas that made OLPC charismatic began cropping up in other projects, making them charismatic in turn. In this way, successive generations of charismatic technologies might all speak to the same cultural narratives, evoke the same social imaginaries, and reflect the same ideologies—just in different forms.

"Charismatic technology" is an analytical anchor for this analysis, one that we will return to throughout the book, along with related theoretical constructs such as the practices of nostalgic design and performing development. Analyzing OLPC through the lens of charisma will make visible the ideological stakes that underpin it and projects like it. We will explore the roots of OLPC's charisma and then what happened to it when its XO laptops were put to use in Paraguay. This journey will help us understand similar projects in development, in education, and across the technology world more generally. Technologists such as those involved with OLPC ignore the origins of charisma at their (and all of our) peril—at the risk of always being bewitched by the next best thing, with little concept of the larger cultural context that technology operates within and little realistic hope for long-term change. Instead, recognizing and critically examining charisma can, as Mosco explains about the sublime, "help us to loosen the powerful grip of myths of the future on the present."[51]

On the flip side, it is just as significant in the story of OLPC and other charismatic technologies to recognize that charisma plays an important, even indispensable, role in our lives. Charisma, whether from leaders or technologies, can provide direction and conviction, smoothing away uncertainties and helping us handle contradictions and adversities. Indeed, since the Industrial Revolution, a faith in technology has played a major role in cultural cohesion. Rob Kling has shown that during the Great Depression of the 1930s, for instance, an "almost blind faith in the power of the machine to make the world a better place helped hold a badly shattered nation together."[52] Contemporary social theorists likewise call on an engagement with fictions, futurisms, and imaginaries of many kinds (whether optimistic or pessimistic) to help chart a course toward a more sustainable and just world.[53]

The trouble comes when these charismatic visions become too removed from the complications of daily life and the arc of history.[54] This happens when we hold a charismatic technology at arm's length, rather than designing it to cohabit the messy world with us and then inhabiting all of its consequences—when we believe that a charismatic technology somehow transcends both historical precedent and everyday life. Mosco has shown that this decontextualized, ahistorical approach was the foundation of much of the misplaced utopianism around the internet.[55] Larry Cuban and David Tyack have shown that the same is true for educational technologies.[56] This book provides a vivid example of what happens when a charismatic technology that was built around particular ideals comes up against the lived realities of its intended beneficiaries—beneficiaries who do not necessarily share those ideals.

As such, though my message may be cautionary, my purpose is not solely to prove charismatic technologies "wrong." Indeed, other accounts already persuasively make similar cases, including Kentaro Toyama's account of "tech solutionism," which critiques the ways in which technology writ broadly is framed as a solution to every problem, and Christo Sims's description of "disruptive fixation," where educational reformers become "fixated" on the promises of technology-driven disruption.[57] These accounts debunk technology-centered hype and can help us understand some aspects of OLPC's story, but this book will also examine what makes a "wrong" charismatic technology continue to be ideologically resonant among some true believers and why specific ideas seem to recur in one charismatic technology after another. Charisma is not inherently good or bad; rather, it is present when a technology's promises outstrip its actual capabilities and capture the social imagination. Whereas a fixation on technology-driven disruption and a strong sense of technological solutionism both play a role in the story of OLPC, charisma provides a way both to unify the various mechanisms for social influence that OLPC tried to harness and to trace their interactions across the world.

Although One Laptop per Child itself may no longer be widely charismatic, some of the ideas that animated the project still are, and they continue to reappear in new projects. Understanding this appeal can help many who engage with technologies—designers and creators, academics, policy makers, educators, development experts, and advocates alike—identify, manage, and even exploit these narratives without getting caught up in them. Through a detailed case study of the consequences of charisma— particularly the performativity of such projects, the treacherous allure of nostalgic design, and the catch-22 that short-term funding and performance expectations create—this book offers those interested in technology-heavy educational reform or development projects a way to recognize charisma, to either harness or resist it. Likewise, it offers technologists a means to critically assess the utopian promises that circulate not just in reform projects such as OLPC but across Silicon Valley more generally.

One Laptop per Child as a Charismatic Technology

From One Laptop per Child's origins to its big debut when Kofi Annan broke off the ultimately doomed hand crank to its realization, distribution, and use in the world, this account will follow the project's laptop from idea to mythology to reality. In the process, it will strive to "stay with the trouble," which, Donna Haraway advises, "requires learning to be truly present, not as a vanishing pivot between awful or edenic pasts and apocalyptic or salvific futures, but as mortal critters entwined in myriad unfinished configurations of places, times, matters, meanings."[58] The following chapters will stay present with the machine, tracing its path from MIT to the Global South, grappling with its messy realities and unfinished configurations along the way.

This book's narrative begins historically. It first examines the intellectual and cultural history of OLPC, which reaches back over fifty years to MIT's nascent hacker culture in the 1960s. It demonstrates the similarities between this hacker culture and the constructionism learning theory that forms a cornerstone of OLPC, showing how both rely on particular social imaginaries about childhood, learning, technology, and schools. From this historical journey, the book then turns to a detailed account of what the project did in the world: how its laptops were used day to day; what the children, teachers, and parents involved with the project thought of them; and what they all learned (as this was meant to be an education project, as Negroponte often claimed).

The first two chapters of this book explore the origins and development of One Laptop per Child, from its precursors at MIT up to the completion

of OLPC's first-generation laptop in 2008. My sources for these (
include speeches about OLPC; discussions on public mailing lists, wiκι,
discussion boards; interviews with some developers; and publications about
constructionism, the educational theory behind OLPC (my empirical and
analytical methods are detailed in Appendix B). In chapter 1, we see that
some of the project's charisma can be explained by the striking similarities
between it and the ethos of MIT's hacker culture—connections that are not
coincidental, as the architects of OLPC counted themselves as members of
this hacker community and credited it with inspiring their work. This chapter then shows why both constructionism and hacker culture continue to
be resonant by unpacking the complementary social imaginaries of school
and childhood, as well as the role that technology is imagined to play in
both. We examine the work that these imaginaries do for projects such as
OLPC by evoking deeply held feelings about the value of (certain types
of) play, creativity, and learning and the kind of child capable of them. In
particular, OLPC implicitly invokes the social imaginary of the *technically
precocious boy*, which developed as an alternative to dominant notions of
masculinity in the United States. This imaginary shows a "natural" mastery
of technical toys as well as a particular kind of rebellious sensibility that
enables technical tinkering—but is still exclusionary by connecting technical prowess to boys in particular. Projects such as OLPC often implicitly
invoke this imaginary, even when trying to be inclusive, when they connect their product to decades of media portrayals and marketing that link
engineering-oriented toys and video games to boys specifically.

In the technology world, these social imaginaries can inspire *nostalgic
design*, and chapter 2 shows how this nostalgia played out in the design
of OLPC's XO laptop. Key features of this laptop—focused on play, freedom, and connectivity—were based on how a number of OLPC developers nostalgically remembered their own (often privileged and idiosyncratic)
childhoods rather than on contemporary childhoods in the Global South.
Moreover, they smoothed away the messiness in their own experiences by
understanding their childhoods through the social imaginary of the technically precocious boy, at times unreflectively treating their experiences as
universal. In doing so, they used the XO laptop's design as a vehicle to evan-
gelize this imaginary to their beneficiaries—specifically, the superiority of a
childhood that is full of pedagogically playful experiences involving technology that would, in particular, be expected to engage clever, scientifically

inclined, and often rebellious boys. Stories of finding escape in computer worlds and of teaching oneself to program in spite of clueless adults played an important role in the project, even as those stories discounted the social and infrastructural resources that enabled this exploration and ultimately perpetuated classed, racial, linguistic, and gendered tropes about what computing culture is and who belongs in it.

Chapters 3 through 6 explore how these ideals played out in practice: how the children using OLPC's laptops day to day made sense of them and how their families, schools, and other institutional and social worlds interacted with them as well. To do this, we will take a close look at an OLPC project in Paraguay. With ten thousand laptops, Paraguay's OLPC project was not the largest. Yet many in the OLPC community considered it to be one of the most successful, as well as one that largely adhered to OLPC's vision—which makes it a particularly interesting test case for the promises that OLPC made.[59] These chapters draw on data collected during seven months of ethnographic fieldwork in Paraguay: six months from mid-June to mid-December 2010 and an additional month of follow-up research in November 2013. This fieldwork was conducted in collaboration with Paraguay Educa, a nonprofit, nongovernmental organization that ran the local project. During this time, I observed children, their teachers, and their families in classrooms, their homes, and public spaces. I typically spent two to three days per week in schools, for a total of fifty-seven school days (or approximately 450 hours), observing classrooms and schoolyards. I conducted 144 interviews with children, parents, teachers, administrators, Paraguay Educa employees, and others involved in the project, nearly all in Spanish.[60] I supplemented this ethnography with some quantitative data, which I use to triangulate results in my analysis. This account connects to the historical thread in the first two chapters to show the material effects of seemingly diffuse ideas, convictions, and values, as they played out in the use of OLPC's laptop.

Drawing on this data, chapter 3 describes the work it took to translate the charisma of OLPC's XO laptops in Paraguay, as well as the individuals and institutions—in particular, teachers and schools—enrolled in that work. In the process, the chapter begins to show how the promises of charisma were brittle in practice, easy to break and difficult to repair. Teachers and students struggled with charging, software management, and breakage. Even though the laptops were built to be both rugged and repairable, in

August 2010 about 15% of students had unusably broken laptops, and more had laptops with missing keys or broken pixels that made them difficult to use. Even when the charismatic potential of the project was translated, there were technical and social barriers to acting on this potential that had been rendered invisible by the stories of easy cultural change this charismatic machine was supposed to enable. These barriers were exemplified by OLPC leaders' comparison of their laptop to a Trojan horse, a device that would give children opportunities to develop into free thinkers independent of, or counter to, the institutions around them.[61] However, the XO laptop was a different kind of Trojan horse in the classroom, one that hijacked lessons with technical difficulties.

In chapter 4, we look at what Paraguayan children with XO laptops did with them in their spare time, as this use case was the cornerstone of OLPC's vision. Here, it will become clear that the social imaginaries that motivated the design and popularization of the XO across the technology world did not translate to most users: fully two-thirds of children hardly ever used their laptops. Some nonuse was due to breakage, which occurred along gendered and socioeconomic lines, complicating some of the benefits the project was supposed to provide. However, most nonuse was due to disinterest: about half of all children in the program in Paraguay were simply uninterested in exploring the laptop on their own time or found it too frustrating to use. This decidedly uncharismatic perspective on OLPC shows that its laptops were not the one-stop liberator that OLPC made them out to be, in the process returning agency to the child beneficiaries of the project. It also demonstrates how the charisma of technologies is limited when its social imaginaries are not shared.

But what about those who *were* using their laptops? Chapter 4 also discusses how most of these children were not interested in learning to program, as OLPC's learning theory had predicted. Instead, they used their machines as portals to music, games, and videos on the internet—even though the design of the laptop made these uses difficult. I find that these media-focused uses were rooted in contemporary computing culture, not in the culture and machines that OLPC employees had used as children. We thus see the limits of charisma based on nostalgic design in resonating with a new generation who grew up with different technical and cultural expectations. Even though I celebrate the agency of these children in finding their own meanings in the laptop, I was also troubled that multinational corporations such as Nickelodeon and Nestlé were eager to capitalize on this new marketing platform to advertise to children via an avenue considered "educational."

Still, there was a very small subset of students—generously, about forty out of the four thousand who had laptops in 2010—who explored the constructionist and creative options of their laptops at least some of the time. Chapter 5 considers the reasons for and implications of such use and what it can tell us about how these children understood the charisma of the machine. Those promoting the charismatic story of laptop-driven transformation in Paraguay framed these students as innately curious learners leapfrogging past teachers, parents, and peers to deeply engage with the machine. In contrast, I found that each student had a constellation of resources that encouraged them down this path: families that steered them toward creative and critical thinking, a focus on the importance of education, and in many cases another computer in the home. Although children's own interests were of course at play as well, their motivation for doing these activities appeared to be less a product of interactions between the universalized child and laptop than a negotiation between many factors, especially parents and peers. This account circumscribes the role that technology itself plays in a child's development, instead highlighting the importance of a child's social worlds in shaping their interests. It moreover shows how the communities in which these children are embedded experience the multiple forms of entrenched marginalization that the laptop was supposed to overcome—particularly gendered marginalization (where only boys were allowed to be "hackers") and linguistic marginalization (where English was uncritically accepted as the universal language of programmers).

Chapter 6 explores the consequences of charismatic technologies for the organizations that build or use them—and specifically the expectation that they perform their charisma. This highlights the catch-22 of charismatic technologies and the projects that support them: they must promise dramatic results to gain social and financial support for reforms, and then they must either admit to not achieving their goals or "perform success" by pretending that they did. This chapter centers on one such performance. Either way, projects that rely on charismatic technologies, such as OLPC, have trouble getting long-term support and, though showy, are often short lived. This catch-22 is pervasive in the technology world and has moreover dogged educational reform for over a century. As the educational technology community moves on to the next charismatic technologies, this catch-22 will continue to hamper the possibility of real, if incremental, change.

Following OLPC and its charismatic machine from its origins through its realization and into the experiences of its beneficiaries may seem overly fixated on the very object I seek to critique. Doesn't such focus, even with a critical eye, risk reifying its charisma? This focus is not idolization, though; it follows a long tradition in STS of tracking a technology to provide a focal point for social critique. "The *object* of the study is not so much to celebrate as to deconstruct the subject," sociologist and STS scholar John Law explains in *The Sociology of Monsters*. The subject, after all, is not a heroic individual (whether human or nonhuman) but "an effect, a product of a set of alliances, of heterogeneous materials." Those who study these technologies—or, in actor–network theory terms, follow the actors—"are not normally guilty of an unexamined and heroic theory of agency," Law says. "Heroes are built out of heterogeneous networks"; they are not born but made by a myriad of influences.[62]

This book, likewise, does not tell the story of OLPC, its XO laptop, or anyone involved with it as an individualistic hero narrative—even though hero narratives are celebrated in the technology world and United States culture more generally. We will see that the project was certainly built out of (and benefited from) heterogeneous networks, and it is the purpose of this book to expose and examine them. We will begin, then, at the beginning, by critically examining the prehistory of One Laptop per Child: the origins, development, and legacy of constructionist learning.

1 OLPC's Charismatic Roots: Constructionism, MIT's Hacker Culture, and the Technically Precocious Boy

The initial ideas [for One Laptop per Child] came in the late 1960s [and] early 1970s, when a man named Seymour Papert made a very simple observation—and that was that children learn different [*sic*] when they write computer programs, because the act of writing a computer program is the closest you can come to thinking about thinking.
—Nicholas Negroponte, quoted in Kleiman, *Web*

We will begin our story with this central question: why did so many so enthusiastically accept One Laptop per Child's charismatic promises? When people imbue technologies with charisma, it is because they expect these technologies not only to be able to solve what they see as problems in the world but to do it in a way that agrees with and amplifies their deeply held core beliefs—their ideological frames—about how the world works. So, in order to understand why OLPC's laptop was charismatic, we need to understand who the developers were, what problems they wanted to solve, and what broader ideological framework they operated within—in short, the project's origin story, as rooted in the individuals, institutions, and imaginaries responsible for the design of OLPC's laptop.

This origin story goes back many years, before One Laptop per Child was announced in 2005 by Nicholas Negroponte, the founder and public face of the project. This story is anchored at the Massachusetts Institute of Technology (MIT); in addition to Negroponte, leadership of OLPC included other professors, researchers, and students from MIT's Media Lab. Walter Bender, a senior research scientist in the Media Lab who had taken over from Negroponte as interim executive director of the Media Lab in 2000, left that position to head up OLPC's software development as vice president of community content (and was reportedly a "close number two" after Negroponte in OLPC's leadership).[1] Mary Lou Jepsen, whom Negroponte recruited to the project when she was interviewing for a faculty position at the Media Lab in early 2005, worked on hardware design as the project's first chief technology officer until she left in early 2008 to found display company Pixel Qi (and was one of the few women on the team, though not often a project spokesperson). Finally, Seymour Papert, who in 2005 was professor emeritus at the MIT Media Lab, had originated the very idea of "one laptop per child" and had discussed how this idea was meant to transform the world.

Papert was an especially important, if often underappreciated, influence on the project. Though OLPC exhibited Negroponte's signature digital utopianism as well, Negroponte readily admitted throughout his marathon of talks in 2005 and 2006 that the very idea for the project was actually Papert's.[2] In these talks, he would often say some version of what he told an audience at the Emerging Technologies Conference at MIT in September 2005: "This [OLPC project] is the life's work of Seymour Papert."[3] OLPC's mission statement likewise credited Papert as inspiring the project. Even with this lineage and Negroponte's consistent crediting, however, Papert's influence on OLPC was still often ignored in the press and even in technology circles. Negroponte was OLPC's undisputed spokesman as he crisscrossed the globe promoting the project. Papert, by contrast, was seventy-six years old and retired when OLPC was first announced in 2005, though he did do some early interviews for the project.[4] This participation was cut short on December 7, 2007—around the time that OLPC's first laptops were coming off Quanta's assembly lines—when Papert was hit by an auto-rickshaw in Vietnam and suffered a brain injury from which he never fully recovered. This further reduced the visibility of his influence.[5]

Those closest to the project, however, knew that its inspiration was Papert and his signature learning theory, *constructionism*. Both were often discussed in project mailing lists and web pages. The OLPC community knew that Papert had written a best-selling book about the idea of one computer per child some thirty-five years before OLPC was announced. Many of them, especially Media Lab affiliates, had been captivated by this book and by previous charismatic projects built by Papert and his students. Thus, in order to understand OLPC's appeal, we must also understand this history. This chapter explores why Papert's ideas remained compelling from when they were first proposed in the 1960s to when they were built into OLPC's laptop in the mid-2000s—in particular, how they paralleled the ethos of the hacker community at MIT and how both tapped into broader social imaginaries of childhood and school that remain popular today. Understanding this lineage is important for understanding the lasting appeal of not just One Laptop per Child but other technology projects in education, development, and beyond that make the same kinds of charismatic promises.

A Brief History of Constructionist Projects

Constructionism's story begins some four decades before One Laptop per Child was publicly announced in 2005. It began when Papert and Negroponte both joined MIT's faculty around the same time in the mid-1960s and became involved in the nascent hacker culture that was developing around the mainframe computers in the university's computer research labs, particularly the lab run by artificial intelligence pioneer Marvin Minsky.[6] Shortly after that, Papert started writing about the potential for children to use computers to learn what he called powerful ideas. Over the next several decades, he elaborated on how these powerful ideas could reshape the lives of children around the world through a learning framework he called constructionism. Many of Papert's projects—spanning some forty-five years of work at MIT, from the 1960s to the late 2000s—express aspects of this learning framework. Though he first began fleshing out details of constructionism in the 1970s in a National Science Foundation grant and a series of working memos, his book *Mindstorms: Children, Computers, and Powerful Ideas* (1980) brought constructionism to a wider audience and remains the core reference on the subject, with a smattering of other books, articles, working memos, and talk transcripts fleshing out additional details.

Mindstorms soon became a best seller—in part because it was published just as personal computers started to gain popularity and the first wave of social anxiety about teaching all children to program was taking hold. But there is also a lot to like in Papert's prose. Witty and engaging, he draws the reader in with a wry and unabashedly personal style. He offers comforting continuity, even as he risks repetitiveness, by returning to the same small set of themes and parables—and the same overall message—across the book and, in fact, across much of his writing and public talks: how would you explain how to make a circle on the ground with your feet? How do you learn how to juggle? How is learning math like learning a language? How is getting to know a new idea like getting to know a person? Why should computers be like pencils? The answers to these questions, Papert argues, hinge on constructionist learning.

Papert's construct*ion*ism borrows heavily from Jean Piaget's theory of construct*iv*ism, as reflected in the confusing similarity between the two names. Before joining MIT in 1964, Papert spent five years at Piaget's International Centre for Genetic Epistemology in Geneva, around the time that Piaget's constructivist theories of child cognitive development—first articulated in the 1920s—started to gain popularity worldwide.[7] Papert adopted from Piaget a focus on children's learning as an active process of constructing knowledge about the world. Both stressed that children (and adults) learn by relating new concepts to what they already know and sometimes adjusting their theories of the world to accept new concepts. Piaget called this "assimilation" and "accommodation," respectively, though Papert called both "Piagetian learning."

At the same time, Papert's constructionism has a number of central elements that diverge from Piaget's constructivism, and describing the foundations of these elements and their implications for One Laptop per Child and Papert's other projects will constitute much of the rest of this chapter. It is also worth noting that Piaget's and Papert's legacies—who reads them and where they have influence—are quite different. Piaget is often still read alongside John Dewey, Lev Vygotsky, and others in schools of education, and the educational research community has amassed some evidence of the value in Piaget's constructivist theories, though with some modifications and critiques along the way. Papert's constructionism, on the other hand, is less commonly taught in education schools, though it remains popular among technology designers.

Mindstorms focuses on describing Papert's most famous application of constructionism, the Logo programming language and its "turtle" interface, which he co-developed with Cynthia Solomon and a number of other students and collaborators from the 1960s through the 1990s. Logo and the turtle were included on OLPC's laptop as the Turtle Art activity. Via the publicity it received from *Mindstorms*—as well as from many hobbyist computing magazines that positioned the programming language as a more fun alternative to BASIC—Logo took the technology world by storm in the early 1980s, much as OLPC would do in 2006 and 2007. The programming language's charisma peaked when it was rolled out in schools nationwide in the mid-1980s—including in my own suburban public elementary school—amid a push to teach programming to young children and a conviction that Logo was the best tool for the job.

In addition to this rollout across the United States, Papert and Negroponte collaborated on building Logo-centric classrooms overseas, starting with Senegal in 1982, where children would learn with Logo on Apple II computers—a plan that Negroponte said was "as audacious as ... putting men on the moon."[8] A February 1984 interview with Papert in *Family Computing* magazine referred to this project when describing Papert as "attempting to cultivate a widespread 'computer culture' especially in Third World societies"—a story repeated for OLPC.[9] In fact, OLPC's website claims this project as a precursor: "In a French government-sponsored pilot project, Papert and Negroponte distribute Apple II microcomputers to school children in a suburb of Dakar, Senegal [in 1982]. The experience confirms one of Papert's central assumptions: children in remote, rural, and poor regions of the world take to computers as easily and naturally as children anywhere."[10] The two installed computer classrooms overseas several more times, including in Costa Rica in 1986 and in Cambodia in 1999, before announcing OLPC in 2005, and Papert advised a one-to-one rollout across the state of Maine.[11] Despite these experiments' varied circumstances and use of off-the-shelf hardware, Negroponte and other OLPC leaders have cited them as the pilot programs for OLPC. "We as a team have been engaged in field studies around the world for almost three decades," states one page on the OLPC Wiki.[12] Negroponte often mentioned Senegal, Costa Rica, and Cambodia as proofs of concept in talks about OLPC.

In constructionism's early days, the education research community was also excited by its promises, and through the 1970s and early 1980s, researchers at Edinburgh University, Kent State University, and the Bank Street College of Education evaluated Logo.[13] However, they found that teaching even the basics of Logo was difficult and that there was scant evidence of skill transfer or the other cognitive benefits that Papert had promised.[14] Meanwhile, the Senegal Logo project fell apart.[15] A 1984 article in *Datamation* magazine reported that the project was too utopian and failed to account for issues such as widespread illiteracy, especially as Logo depended on written English. It also questioned the project's motives as colonialist: "Perhaps the most damaging question raised about [the center] concerns the conviction that helping Third World countries acquire computer technologies would be beneficial. To many critics, such a goal is an artifact of colonialism, imposing Western values and definitions of progress on other cultures for less than altruistic reasons."[16] Such critiques would resurface in discussions of OLPC.

As these results percolated through the enthusiast community and as the charisma of Logo started to wear thin with the struggles of day-to-day use, others stepped forward to voice frustrations with Logo's failure to live up to its charismatic promises. One frequent columnist in *Compute!* magazine described how it was easy to master Logo's basic commands but nearly impossible for new programmers to jump from there to recursion or other more advanced features. Those who did manage the jump found Logo—an interpreted language, rather than a compiled one—hopelessly slow and a hog of precious (and very expensive) memory. The columnist concluded, "Five years ago I predicted the demise of BASIC and its eventual displacement by Logo as a programming language for neophytes. As I look back on the past five years, I see that my own vision was clouded by my enthusiasm and that what I saw was largely a dream, not an accurate reflection of the world of educational computing."[17]

Papert's Logo group saw these critiques—but Papert dug in his heels. Although he acknowledged that "Logo is notoriously slow ... [and] is seen as a language for 'toy' programs that may use interesting ideas but do not do useful work," he brushes off these critiques, stating that when another team chose "speed over mathematical transparency" in an alternate implementation of LOGO, they were not "nurturing a 'mathematical aesthetic.'"[18] In the same essay, he also brushed off the Bank Street College of Education study as fatally flawed, "'obviously' absurd," a "very slim experiment," and "a self-defeating parody of scientism."[19] He disparaged one of the coauthors of the Bank Street study specifically, concluding that "Dr. Pea's criterion for how Logo is supposed to improve thinking skills ... is a good example of the conservatism inherent in traditional experimental methodology" and that he "cannot see how anything useful can be derived from the Bank Street finding that the children did not meet Dr. Pea's criteria of planning." These kinds of positions were not new: many of Papert's books and papers disparaged educational experiments and the researchers carrying them out. In the preface to *Mindstorms*, Papert stated, "If any 'scientific' educational psychologist had tried to 'measure' the effects of this encounter, he would probably have failed. It had profound consequences but, I conjecture, only very many years later. A 'pre- and post-' test at age two would have missed them."[20] In subsequent writing, he lumped evaluators and "white-coated academics" in with all that was wrong with school.[21]

As this tumult was unfolding, Negroponte was working with fellow MIT professor Jerome Wiesner, Papert, and others to establish the MIT Media Lab. When it opened in 1985, it embraced constructionism and Papert—despite the critiques that were brewing. From the beginning, *Mindstorms* was woven into the school's culture as a core text of the Program in Media Arts and Sciences graduate course Technologies for Creative Learning (later renamed Learning Creative Learning). And even when Logo had been largely abandoned as a national project, constructionism's charisma lived on in the Media Lab and continued to inspire projects throughout the 1990s and 2000s, including Lego Mindstorms, FabLab makerspaces, the Phidgets and Makey Makey prototyping kits, and the Scratch programming environment.[22]

At the same time, the broader community interested in construction-ist approaches did evolve this learning theory in response to the evalua-tions throughout the 1980s and beyond, generally taking a much more pedagogically-focused approach than Papert originally called for. For instance, Richard Noss and Celia Hoyles's 1996 book, *Windows on Math-ematical Meanings*, provides a road map for educators and researchers interested in using some elements of constructionism for mathematical education, though with some modifications and a firmer grounding in other educational literature. Though *Mindstorms* has more than ten times the number of citations than *Windows on Mathematical Meanings*, the latter has still been influential in the broader field of educational technology. In an endorsement printed on the back on the book, Papert himself praised it as "the most elaborated theoretical discussion to date" of using comput-ers for mathematical education. A number of other scholars have contin-ued evolving constructionism as well, though their audience has largely remained educators and education researchers, rather than the technology design world more broadly. One exception is MIT professor Mitch Resnick, who is in many ways Papert's successor as the LEGO Papert Professor of Learning Research and the director of the Lifelong Kindergarten Group at the MIT Media Lab. In his work co-developing the Scratch programming environment, Resnick has shifted the interpretation of constructionism used for Scratch to accommodate some of the critiques of it.[23]

Some threads of constructionist thought, however, have persisted relatively unchanged from Papert's initial configuration—particularly those promoted by some of the figures who would go on to found OLPC. In 1996, the same year that *Windows on Mathematical Meanings* was published, Papert's book *The Connected Family* and the paperback edition of Negroponte's *Being Digital* also came out, both of them retrenching the vision of constructionism put forth in *Mindstorms*. *Being Digital* included a chapter that praised Papert's work and recast the failed experiment in Senegal as a success; it stated, "The children from this rural, poor, and underdeveloped west African nation dove into these computers with the same ease and abandon as any child from middle-class, suburban America."[24] Even more importantly here, it was Papert's original vision for constructionism that provided the foundation for One Laptop per Child. For this reason, when I refer to "constructionism" in this book, I refer specifically to this original vision put forth by Papert, rather than the broader community involved with subsequent constructionist learning research.

Constructionism's Hacker Origins

What explains Papert's retrenchment, as well as the persistent charisma of constructionism in spite of its tumultuous history? The answer has to do with the parallels between constructionism and the computer-centered hacker culture at MIT in which Negroponte and especially Papert found camaraderie and inspiration—and the underlying social imaginaries that made both constructionism and hacker culture resonate then and continue to resonate now across the technology world.

Today's dominant narrative of technological protectionism, often discussed in the United States as the need to limit "screen time," projects technology as a danger to what is most valuable about childhood.[25] But a strong counternarrative frames technology instead as a contributor to playful exploration and creativity, proclaiming the potential for children to learn to use technology in transformative ways. These narratives, even when seemingly opposed, share common social imaginaries of childhood as a more noble state and of the best childhoods as marked by particular kinds of creative, rebellious, entrepreneurial play—and in opposition to school, at least as it is commonly conceptualized.

These social imaginaries—which circulate through media represen-tations, shape conversations about childhood and children, and affect people's sense of identity—can also influence technology worlds and the projects they create. In particular, MIT's early hacker culture played an important role in shaping the aspects of Papert's constructionism that are distinct from Piaget's constructivism—and both the hacker community and constructionism presuppose the social imaginary of the technically preco-cious boy. This imaginary draws on related imaginaries of the innate tech-nical creativity and rebellious nature of children (especially boys) and the stultifying effects of "school as factory" as a target for this rebellion. Its reso-nance explains why successive constructionist projects—from Logo to Lego Mindstorms to Scratch to One Laptop per Child—have all been charismatic to many in the technology world and beyond.

1. Freedom and the "Proteus of Machines"

One major difference between Piaget and Papert has to do with how Pap-ert makes computers central to learning. Observing that the process of knowledge construction among young children is often embodied—that is, related to our own bodies and other physical objects around us—Papert emphasizes what he calls "body knowledge" and manipulation of "materi-als" as central to learning much more than Piaget does.[26] Papert calls materi-als that are particularly useful for learning "objects-to-think-with."[27] Using an object-to-think-with, Papert argues, helps children relate to what they are learning physically or sensorially. These objects-to-think-with should be able to grow with the child, enabling them to continue to think about increasingly complex and abstract ideas in terms of concrete (or at least visual) affordances.

Taking this one step further, Papert argues that one of the best objects-to-think-with is a programmable computer. Describing the computer as "the Proteus of machines" (in reference to the Greek sea god who could change his shape at will), Papert says it can appeal to many audiences and exert a "holding power" over children.[28] He extols the educational potential of programming and advocates giving children unrestricted access to com-puters so they can tinker as much as they want. In *Mindstorms*, he writes, "My vision of a new kind of learning environment demands free contact between children and computers."[29] Children should be able to access all aspects of the computer and have the freedom to dive as deeply as they wish into its inner workings. Papert asserts that when given unlimited access to a computer, children can also be empowered to think "like a computer," learning the "language" of the machine just as living in another country when young would enable them to learn the language spoken there with a proficiency that adult learners can only envy. Under these conditions, Papert says, computers could finally scale up the ideals of "progressive education."[30]

Papert attributes his realization that computers can be powerful tools for learning to his experiences with MIT's hacker community, which he first encountered in Minsky's Artificial Intelligence Lab in the 1960s. This group, which was first described in detail by journalist Steven Levy in his book *Hackers: Heroes of the Computer Revolution* (1984) and later analyzed by scholars tracing early computing cultures, also believed in unrestricted access to the mainframe computers in Minsky's lab—Papert describes joining this group for all-night hacking sessions. Extending MIT's longstanding culture of elaborate "hacks" (pranks), this group started calling themselves "hackers." Far from the nefarious connotations the term has today among the general population, these men (and, in Levy's account, they were all male) focused on playful computer use and deep understanding. Although they may have viewed rules and conventions (such as computer lab open hours) as something that could perhaps be bent or ignored, this was not the defining characteristic of the group—their technical passion was.[31] The learning they did on those mainframes was the same kind of hands-on learning that constructionism advocates, and Papert argues that those early computers also exhibited the "low floor" and "no ceiling" characteristics of the best objects-to-think-with. Thus, one of the central tenets of Papert's constructionism—that children should have unrestricted access to computers—originated not with Piaget but with MIT's hacker culture.

2. Yearners, Passion, and Creative Play

Moreover, this community of hackers exhibited the passion and play that Papert later made cornerstones of constructionist learning. Papert explains that if children learn via constructionist principles, they will maintain a love and excitement for learning, and that this affective component is crucial to constructionism.[32] Papert encapsulates this passion and playfulness in a prototypical child that he called the *yearner*. In his 1996 book, *The Children's Machine*, he explains that children are born yearners and that some adults are able to retain that childlike state throughout their lives—in effect to "never grow up," as expressed in *Peter Pan*. A key element to Papert's description of yearners is that the state is innate: thinking independently, not caring about others' opinions, and seeking answers via many routes for questions in which they are personally interested are not skills that are learned; they are inborn and only unlearned. "Children seem to be innately gifted learners," Papert explains in *Mindstorms*, "acquiring long before they go to school a vast quantity of knowledge by a process called 'Piagetian learning,' or 'learning without being taught.'"[33]

This passion is something Papert experienced himself in his early days at MIT. In *The Children's Machine*, he explains, "Many factors made the move [to MIT] attractive, ... [such as] a wonderful sense of playfulness that I had experienced there on brief visits. When I finally arrived, all this came together in all-night sessions around a PDP-1 computer that had been given to Minsky. It was pure play." He goes on to describe his experiences with the hackers in Minsky's lab as "playing like a child and experiencing a volcanic explosion of creativity."[34] Papert wanted children around the world to have this same feeling: "I had my first experience of the excitement and the holding power that keeps people working all night with their computers. I realized that children might be able to enjoy the same advantages—a thought that changed my life."[35]

Another part of constructionism that differs from Piaget's constructivism—and bears a clear relationship to hacker practices—is its focus on debugging. In constructionism, this refers to the process of embracing "wrong" ideas and then iteratively revising them by testing.[36] In computer programming, it refers to the process of finding and correcting the mistakes (or bugs) in computer code.[37] Papert argues that an openness to debugging is among the most important aspects of constructionism: it encourages children not to internalize feelings of failure when they have a "wrong" answer but to see this as an integral part of the learning process.[38]

Papert's yearners and the playfulness they exhibit connect with a particular social imaginary of childhood that is characterized by seemingly innate creativity, fearlessness (whether with physical feats or technology), innocent mischief, and "rugged individualism"—one that is in close communion with the natural world but utterly out of sync with the adult world.[39] This imaginary seems so natural to many, especially in the United States, that it is often assumed to be timeless and universal; it has been a constant refrain in English and American media for more than 150 years, evinced by the perennial resonance and many variations of characters such as Huckleberry Finn and Peter Pan.[40] Within this cultural frame, it seems self-evident that childhood is purer and nobler than the cluttered, compromised, and culture-bound realities of adulthood—that children are in effect outside of culture, and only as they age do they become entangled in cultural logics and concerns.[41]

Although this imaginary is generally perceived as pervasive and timeless, its provenance is relatively recent. The romantic notions of childhood that are common today are a product of post-Enlightenment shifts in thinking and in particular of the worldview of the Romantic era, a set of nineteenth-century ideological shifts in the United States and western Europe that coincided with the Industrial Revolution.[42] The same reform-minded movements that aimed to take children out of factories and put them into schools also helped redefine childhood as the distinct developmental stage that we understand it to be today: a stage characterized by innocence that should be protected to enable play, exploration, and learning.[43]

After World War II, as a generation of Americans settled in newly built suburbs across the United States with the help of the GI Bill and sent children to newly built public schools, this imaginary narrowed to focus on individualistic, entrepreneurial creativity, enabled by an unprecedented number of IQ-boosting, creativity-enhancing toys—toys that were particularly marketed toward boys as the stewards of the country's economic future.[44] The results of this shift continue to be felt today: the dominant American imaginary of childhood (and especially for white, middle-class boys as the archetypal American child) continues to be nearly synonymous with the creativity and play that these open-ended toys are designed to elicit. The imaginary of the naturally creative child de-emphasizes the extent to which creativity and play are socially motivated and socially learned. In particular, it ignores the important role that parents and other adults play in structuring a child's world—by providing toys, environments, activities, role models, and more—to elicit certain kinds of play and creativity.

In Papert's thinking, computers are the epitome of these open-ended toys. Like the others in Minsky's hacker group, Papert argues that computers especially can inspire passion and play. Part of this captivation is the magic and power that result from understanding programming concepts enough to command the machine; as Papert often liked to repeat in his writing, his goal with constructionism is to have the child program the computer, not be programmed by it.[45] But unlike the other hackers in Minsky's lab—who seemed to revel in the knowledge that their idiosyncratic passion for computers was not shared by most—Papert argues that given unlimited access to a computer early enough, *all* children will love the machine and that this love will change their lives for the better. In *Mindstorms*, Papert writes, "I am essentially optimistic—some might say utopian—about the effect of computers on society. ... I too see the computer presence as a potent influence on the human mind."[46]

This stance is less universally accepted. Even though the social imaginary of the naturally creative child is fairly uncontested, ideas about the role that technology might play within it are not. At times childhood is portrayed as fragile, and technology as a corrupting influence on it. Television, video games, the internet, social media, and mobile phones all threaten its innocence with the encroachment of adult concerns: popularity, consumerism, sexualization (especially for girls), violence (especially for boys), and other forms of "growing up too fast." The American popular press employs the pathological language of addiction, toxicity, and contagion and narratives such as "screen time" to foment a culture of fear when it comes to children using technology.

In Papert's constructionism, however, technology is portrayed as an important part of childhood, and its mastery is folded into the imaginary of the naturally creative child. Indeed, many who have watched a child with a touch screen or a video game have marveled at how children seem to be naturally enamored with technology (which, of course, is influenced by adults being enamored with it as well—but that social motivation is often ignored). Negroponte and Papert both repeatedly lean on the common trope of precocious young children, especially boys, who seem to take to electronics fearlessly and naturally—a counterimaginary which, far from being stunted by technology, experiences power and freedom in technical mastery.[47] Many technical communities, such as OLPC and the MIT Media Lab from whence OLPC came, align themselves with the latter discourse. They assert that technology, from radio to computers, can foster children's creativity rather than squelch it. The imaginary that undergirds this view is what I will call the "technically precocious boy," which I will explore in more detail below.

Papert moreover argues that understanding programming concepts will unlock the keys to understanding one's own learning processes—an idea that originated in the field into which he was hired at MIT: artificial intelligence. In the 1960s, artificial intelligence had become the epicenter of the diverse and exciting field of cybernetics, and MIT was the epicenter of this community. Attracting some of the biggest names in science to wide-ranging conversations, cybernetics was a continuation of some of the utopian thinking about machines that first developed around World War II and animated various research initiatives funded by the Advanced Research Projects Agency (or ARPA, with "Defense" added to the beginning of the name in 1972), including Minsky's lab.[48]

Papert describes cybernetics rapturously in *Mindstorms* as a potent framework for understanding learning by thinking about brains as we think about computers.[49] This was not unusual at the time. There was a widespread belief that human brains are particularly sophisticated computers; Minsky, in fact, famously called them "meat machines." Although the problems with this equivalence eventually contributed to the collapse of cybernetics and to the "artificial intelligence winter" during which funding was scarce and the field stagnated for decades, fragments of the belief persisted—including a faith in computer-enabled constructionism as a pathway to learning how to learn, which also animated One Laptop per Child.[50]

3. The School as Factory

In contrast to the playful joy of learning and debugging on a computer, Papert claims that school is an environment that encourages children to internalize feelings of failure. Even though Papert does not explore what makes kids resistant to getting things "wrong," he squarely blames school culture for causing the shame they might experience when it happens.[51] Throughout *Mindstorms* and even more ardently in *The Children's Machine*, Papert equates school with the worst kind of disembodied rote learning. "School," in Papert's writing, is a monolith of creativity-squashing drill and test, of unintuitive facts and incomprehensible equations, and, above all, of "dishonest" "double-talk."[52] The classroom is "an artificial and inefficient learning environment that society has been forced to invent because its informal environments fail."[53] Moreover, he writes, "School has an inherent tendency to infantilize children by placing them in a position of having to do as they are told, to occupy themselves with work dictated by someone else and that, moreover, has no intrinsic value—school-work is done only because the designer of a curriculum decided that doing the work would shape the doer in a desirable form."[54]

In schools, Papert explains, "[e]ducators distort Piaget's message by see-ing his contribution as revealing that children hold false beliefs, which they, the educators, must overcome. ... Children are being force-fed 'cor-rect' theories before they are ready to invent them."[55] Removed from chil-dren's physical worlds, school does not involve play and does not naturally build on children's existing mental models; it "rejects the 'false theories' of children, thereby rejecting the way children really learn."[56] Papert wonders, "Why, through a period when so much human activity has been revolu-tionized, have we not seen comparable change in the way we help our children learn?"[57]

While Papert is unequivocal about his disdain for school, the role of teachers in constructionist learning is more contradictory. Throughout his writing, he says that children learn "spontaneously" and "without being taught" when outside the classroom, at least as it is currently formulated.[58] At the same time, he provides advice for teachers to incorporate construc-tionism into the classroom, including ideas for lessons. He depicts adults as hopelessly mired in petty social norms and lampoons the ways that they impose on children what they have decided are the right ways of think-ing.[59] Yet, he then describes how he uses constructionist tools to introduce children to concepts that *he* considers important to learn, such as recur-sion and combinatorics. More in line with Piaget's constructivism, Papert supports the idea of casting teachers as co-learners.[60] For example, teachers can pursue a goal that they do not know how to accomplish either and puzzle it out alongside students, creating an empowering, authentic learn-ing experience.[61]

One Laptop per Child, on the other hand, was clearer in its messages about teachers, especially in the early years of the project. While Negro-ponte's tagline that OLPC was "an education project, not a laptop project" may seem supportive of education, OLPC leadership often made statements that were both anti-school and anti-teacher.[62] At a press conference in 2006, Negroponte said, "Now when you go to these rural schools, the teacher can be very well meaning, but the teacher might only have a sixth-grade education. In some countries, which I'll leave unnamed, as many of as one-third of the teachers never show up at school. And some percent show up drunk."[63] In a 2007 interview, Walter Bender echoed these claims. "So this is who we're developing the laptop for," Bender said. "There are one billion school-age children in the developing world. A lot of these kids' schools might be under a tree; they're lucky if the teacher shows up."[64] Later, in a 2013 book defending OLPC, Bender once again restated the project's goals of being an all-in-one solution: "My personal focus is giving children the tools they need to make learning happen, even in absence of an effective, formalized school experience."[65] These sentiments were not only directed at the Global South. In his book *Being Digital*, Negroponte disparaged his own (expensive private-school) education and talked of making MIT's Media Lab a haven for others who, like him, struggled with school.[66]

One would think that such dim views of an institution that ostensibly has similar goals—helping children learn—would be incongruous, to say the least. That, however, was not the reaction Papert or OLPC received. Instead, in academic critiques and news articles alike, these condemnations were rarely questioned or even noted. Similar sentiments are certainly commonplace across the technology world, with educational technology companies often assuming a problematic public-school system in desperate need of technology-enabled reconstruction or even dismantling.[67] Moreover, these anti-school opinions are shared far beyond the technology world; in the United States especially, they are ubiquitous enough to feel mundane. They reflect a broader American social imaginary that school is a stultifying institution virtually unchanged for well over one hundred years.

Much of this social imaginary can be summed up by the metaphorical "factory model" of education, to which Papert refers obliquely in his writing.[68] While references to the factory model go back to the late nineteenth century (and was evoked by the band Pink Floyd in their 1979 rock opera *The Wall*), the term's use skyrocketed in the 1990s, as public discourse about education became dominated by hand-wringing about slipping national competitiveness and a lack of twenty-first-century skills.[69] Depending on who is invoking it, this *school as factory* might prepare children for being factory workers—cogs in a capitalist machine—or it might employ factory-like methods in education (or both). Either way, the implication is negative, as it erases the skill of teachers, the humanity of students, and the agency of all involved behind dystopian connotations of automation and assembly lines. When yearners have tried to innovate within schools, Papert says in *Mindstorms*, they are "driven out in frustration."[70]

However, Papert says there is a "weak link in the vicious cycle" of institutionalized education: children's love of computers. Computers can "so modify the learning environment outside the classrooms that much if not all the knowledge schools presently try to teach with such pain and expense and such limited success will be learned, as the child learns to talk, painlessly, successfully, and without organized instruction." He goes on, "This obviously implies that schools as we know them today will have no place in the future. But it is an open question whether they will adapt by transforming themselves into something new or wither away and be replaced."[71] Moreover, Papert argues that computers uniquely enable these possibilities; it is only with the rise of personal computers that personalized and playful learning became possible. In a televised talk in the mid-1980s (the transcript of which he posted on his website), Papert explained, "I think that the key point about this technology and education is that ... it enables us to not put children through that traumatic and dangerous and precarious process of schooling." He continued,

> Nothing is more ridiculous than the idea that this technology can be used to *improve* school. It's going to *displace* school and the way we have understood school. ... What's wrong with school is absolutely fundamental. ... School means a place where children are segregated from society and segregated among themselves by age and put through a curriculum. ... I'm saying that it is inconceivable that school as we've known it will continue. Inconceivable. And the reason why it's inconceivable is that little glimmer with my grandson who is used to finding knowledge when he wants to and can get it when he needs it [on a computer], and can get in touch with other people and teachers, not because they are appointed by the state, but because he can contact them in some network somewhere.[72]

It is worth noting that the social imaginary of school as factory, for all its rhetorical power, never accurately described the state of education.[73] Indeed, historians have shown that this factory model was just one among a steady stream of educational reforms from the first calls for universal public schooling.[74] The incongruities between the reality of a school system that is constantly "tinkering" with reforms and the imaginary of an unchanged and mechanical factory invite the question: why has the social imaginary of the factory model persisted? It turns out that it does cultural work for a certain group of powerful actors. For one, it renders the educational system both in urgent need of fixing and receptive to quick fixes—a site that matches utopian educational reform narratives such as Papert's and OLPC's. "Impatient with the glacial pace of incremental reform, free of institutional memories of past shooting-star reforms" that failed to revolutionize day-to-day schooling and "sometimes hoping for quick profits as well as a quick fix," educational historians David Tyack and Larry Cuban explain, reformers point to technology—such as Papert's Logo-enabled computers and OLPC's laptops—as the best method to "reinvent" education.[75]

4. Antiauthority and the Glorification of Healthy Rebellion

The deep distrust of school apparent in these descriptions of construction-ism reflects the hacker community's mistrust of authority figures more gen-erally—a mistrust so codified in the culture that Steven Levy canonized it in two of the six tenets of what he called the "hacker ethic." Levy described this ethic, the core principles of the hacker community, drawing on the time he spent with the hacker group at MIT; the ethic was subsequently embraced by many in the hacker world. The third tenet reads, "Mistrust authority; promote decentralization," and the fourth is "Hackers should be judged by their hacking, not bogus criteria such as degrees, age, race, sex, or position."[76] This group decried school as an example of the authority they distrusted; instead, they found great solace in computers and celebrated others who felt the same. Computers provided these odd characters with a common topic to bond over. Levy describes how even before the days of computer networks, the computer provided a central point of contact for the community and that it did not matter if you were a professor or just some "guy" (as Levy repeatedly calls this group of all men) off the street, as long as you could code.[77] Moreover, these hackers described learning about computers in terms of passion and freedom, contrasted with the bor-ing, stifling, and unfulfilling classroom—sentiments that Papert echoed in his descriptions of constructionism.[78] Even though not all in the commu-nity actually rejected school—to be graduate students or professors at MIT, many had excelled—they still upheld this antiestablishment stance, and it continued to be important in the software development world, making Papert's messages remain resonant.[79]

This antiauthority stance connects to broader American social imagi-naries about the role that rebellion in childhood—and especially in boyhood—plays in influencing what kinds of behavior adults tolerate and what kinds of technical pastimes they make available. A certain degree of "healthy rebellion" has even been encouraged as integral to American boy culture.[80] Far from the more ideologically intimidating rebellion that actu-ally threatens to change the status quo, the kind of healthy rebellion that is sanctioned in this social imaginary is often tolerated as "boys will be boys"—with (white, middle-class) boys as the presumed rebels. At times it is even encouraged as freethinking individualism. From Mark Twain's *Tom Sawyer* to today, popular culture has linked healthy rebellion with creative confidence, driven by naturally oppositional masculine sensibilities—the kind that helps produce ruthless entrepreneurs and brilliant businessmen. This form of rebellion has even come to be seen as a natural part of boys' play that is worth celebrating and, if possible, rechanneling into productive ends, encouraging children to develop useful skills even as they satisfy their desires to slip the shackles of adult supervision, adult-approved activities, and adult-defined rules.[81]

This desire to both encourage and channel rebellion in childhood has influenced what technologies adults make available to children—and which children they focus on. Technical books such as *The Boy Mechanic: 700 Things for Boys to Do* (1913), the frontispiece of which "shows a boy ready to step off a cliff in a glider that he has built according to the instructions on page 171," provided one avenue to channel this rebellion and a clearly-stated target: boys.[82] Later, certain kinds of masculine rebellion became an integral part of early cyberculture communities, which historians Fred Turner and Nathan Ensmenger have shown adopted individualist countercultural norms to push back against early notions of computer programmers as mindless suits, low-status clerical work, or tools for corporate control.[83] Narratives about the kinds of "all in good fun" masculine rebellion that computers could enable were popularized in the 1980s through nonfiction books such as Steven Levy's *Hackers*, novels such as *Neuromancer* by William Gibson, and movies such as *War Games* and *Tron*.[84] These media imbued cyberspace with metaphors steeped in rebellion: of the Wild West, of Manifest Destiny, of a new frontier of radical individualism and ecstatic self-fulfillment.[85]

It also encouraged a libertarian sensibility, where each actor was considered responsible for their own actions, education, and livelihoods—even when that meant ignoring the massive infrastructures that made the seemingly freewheeling space of the cyberspace frontier possible. The resulting illusions included newfound freedoms and potentials for computer-based self-governance, the inversion of traditional social institutions (putting, of course, computer-savvy hackers at the top), the flattening of bureaucracies, the end of geographic inequity, and a reboot of history.[86] Even though computerization has largely entrenched existing power structures, these ideals of the computer age live on, especially among technologists who have not had much contact with those actively excluded from the technology world.

Not coincidentally, this era of Gibson's law-flouting "console cowboys" and Levy's youthful hacker pranks occurred during the childhoods or early adulthoods of many of those involved with One Laptop per Child, and some, such as Papert and Negroponte, write about being active participants in the early stages of this world. The imaginary of programming computers as a mode of masculine rebellion—whether against school, peers, or societal norms—was also common among OLPC's developers. For instance, Benjamin "Mako" Hill, a longtime advisor to OLPC with a PhD from the MIT Media Lab, has described with great candor his educational journey through an attention deficit disorder diagnosis, Ritalin, and various public and private schools in an online essay titled "The Geek Shall Inherit the Earth: My Story of Unlearning." He writes about how strongly he identified with "The Hacker Manifesto"—first published as "The Conscience of a Hacker" in 1986 in the hacker magazine *Phrack*—when he encountered it as an alienated teen. In his essay, he mimics its fierce and antagonistic, yet also eloquent, description of the antagonism of school and the camaraderie of the computer, also (quite possibly intentionally) using some of the same language as Papert about school "molding" children versus letting them develop freely: "By day, I felt school forcing me into a rigid and uncomfortable mold—often resorting to chemical means to accomplish the feat. By night, I was able to learn, build, explore, create, and expand myself, both socially and educationally—an ability only afforded to me through my use of technology."[87]

This story and others that OLPC contributors told about the potential of computers for children roll into one package many of the same promises connected to computers and cyberspace as sites for creative play and rebellion, places where yearners will find a home. Framing these oppositional attitudes as natural and good, Papert's and Negroponte's writing often praised iconoclastic or freethinking children who took to computers (and to their experiments) easily, glossing over the many complexities of childhood by universalizing children as natural yearners while simultaneously signaling the technical interests they expected yearners to have.

The kinds of learning these children would supposedly accomplish as a result may well fly in the face of the kinds of learning that teachers, parents, or other adults may want. However, Negroponte and other OLPC leaders suggested that such rebellion should be encouraged as the expressions of yearners against school and stifling societal norms—norms that, in their view, had even held back the Global South on the world stage. After all, Papert claimed, millions of people around the world learned how to use and even program computers in spite of schools—indeed, "without any-body teaching them."[88] With computers, he claimed, this rebellion against school in particular could finally be realized; students would become per-fect neoliberal subjects, determining their own learning path without inter-ference from institutionalized education or the state. As he explained in *Mindstorms*, "Education will become more of a private act, and people with good ideas ... will be able to offer them in an open marketplace directly to consumers."[89]

Gender and the Technically Precocious Boy

This shared ethos between constructionism and hacker culture at MIT and beyond—a belief in the universal appeal of computers, in the natural cre-ativity of children, in the detrimental effects of school, and in the value of certain kinds of rebellion—are all aspects of a particular social imaginary of childhood that is widely recognized across American culture: that of the *technically precocious boy*. While common enough (and often enough por-trayed in media), this imaginary is generally defined in opposition to the dominant imaginary of boyhood, which we might call the "jock imagi-nary." This jock imaginary is characterized by macho athleticism and often a rejection of more intellectual pursuits as not masculine enough. However, the technically precocious boy and the jock imaginaries are not as diametri-cally opposed as they may seem; they share a culture of competitiveness and sometimes physicality, even though the competition is turned to dif-ferent ends. Instead of feats of strength or skill in sports, the technically precocious boy shows off feats of technical prowess and ingenuity—what the MIT computer culture in which Papert and Negroponte participated would call "hacks." More importantly, these two imaginaries share an exclusionary masculinity that is often hostile to outsiders, including girls and women, people of color, and others who do not fit the gendered norms implicit in the imaginaries, as we will see.

The technical toys that support the social imaginary of the technically precocious boy have helped define American boyhood in particular for over a century and have been lauded for supporting their target audience's supposedly natural inclination to tinker and explore. While individuals of course have the agency to choose whether to accept or reject cultural norms, the ways that various technologies—from radios to transistors, from electric trains to computers—have signaled to boys and invoked boyhood have had significant cultural effects. This association is as old as the imaginary of childhood more generally—and likewise connects these boys to a more primitive, natural state, which Amy Ogata calls the "savage boy inventor."[90] The association intensified as the mass-produced toy industry found its footing after World War I and took off after World War II.[91] Starting in the 1920s, toy manufacturers and their catalogues, instructional manuals, radio (and later television) shows and commercials, and national contests reinforced the still-new idea that engineering was a space for natural masculine creativity, and its ideal protégé was the technically precocious boy.[92]

Girls were excluded from this imaginary from its inception. For one, the marketing of technical toys and the manuals that accompanied them suggested (often through language and by omission) or even stated outright that girls were unsuited to these kinds of technical pursuits. Like the association between boys and outdoor play or boys and intellectual achievement, however, this connection was socially constructed, and the association was both new and far from universal. The association did additional important work in making the various rapidly growing fields of engineering more attractive for boys; if they identified with the imaginary of the technically precocious boy, they would find an outlet for this identity and kindred spirits there.[93] Over time, as boys solidified parts of their own identities around this imaginary, it became reified, its tenets reinforced, and its basis seemingly even more deeply entrenched in boyhood.

The connection between this imaginary of the technically precocious boy and computers in particular is more recent. Before the 1960s, computer programming—even though it required sophisticated problem-solving and complex knowledge—was generally viewed as a low-status, clerical profession separate from engineering, and was filled with relatively poorly paid women. But starting in the 1960s and continuing in the 1970s and 1980s, a growing computer industry—as well as dedicated hobbyists and other electronics enthusiasts and the magazines that sustained their communities—worked to shift this perception to align itself with other engineering disciplines, and they co-opted the social imaginary of the technically precocious boy to help them do so.[94]

As part of MIT's hacker group in the 1960s and 1970s, Papert was part of this movement to rehabilitate programming, advocating constructionism as a pathway to joyful technical mastery in the same way that tinkering with technical toys did for previous generations (including his own). Taking up Papert's book *Mindstorms* with gusto, the largely masculine hobbyist computing community was then integral in popularizing constructionist learning theory in the 1980s. Video games—which had shifted from a niche hobby to a boy-focused media empire in the mid-1980s—also became implicated in this imaginary, and Papert lauded video games for subverting traditional education. Teachers, he said, should be "struck by the level of intellectual effort that the children were putting into this activity [of playing video games] and the level of learning that was taking place, a level that seemed far beyond that which had taken place just a few hours earlier in school."[95] The precision and technical mastery required by boys' model-building contests in the early to mid-twentieth century were transferred onto these virtual games, and their guns, cars, planes, and physical competitions reflected the concerns and sensibilities of those boyhood worlds—and almost always featured a male protagonist.[96]

In the last decade, toy and technology manufacturers in the United States and around the world have broadened their markets and have ostensibly started to target some toys toward a more gender-neutral audience. However, without efforts to actively and specifically push back on the effects of decades of gendered targeting, this legacy lives on: the boy user for these products may no longer be explicit, but he is still implied—what scholars in cultural studies would call "unmarked."[97] Likewise, the online world heralded as gender egalitarian in the heady days of cyberutopianism in the 1980s and 1990s has also proven to be a sexist (as well as deeply racist) space, where those who actively identify as anything other than white boys or men—or who take up social positions that question the dominance of this social group—in public online spaces are often explicitly harassed and excluded.[98]

With its focus on programming and technical tinkering, One Laptop per Child likewise fit into this tacitly gendered narrative. Indeed, its charisma relied on its connection with this social imaginary, which continues to be a powerful force in American culture. While the programming-oriented tinkering that OLPC's leaders promoted was not something that any of them said only boys could or would do, without specifically pushing back against the decades spent marketing technical toys and programming to boys in particular, OLPC fell back on the imaginary of the technically precocious boy that similar products have more explicitly targeted. This implicit bias is also reflected in Papert's writings on constructionism, which are peppered with examples of technically precocious boys enthusiastically engaging with his programs.

At the same time, Papert did not specifically exclude girls; his writing includes occasional examples of girls and speaks in general terms about "children." *Mindstorms* and Papert's other publications on constructionism discuss how programming can be "artistic," like writing poetry. On the one hand, this echoes the values of the male-dominated hacker group of which he counted himself a member and was even mentioned in Levy's hacker ethic. On the other hand, this could open up the field to different ways of knowing. Papert and his third wife, fellow MIT professor Sherry Turkle, explored this in relation to gender specifically. Their article "Epistemological Pluralism" attempts to separate computer programming skills from what they acknowledge is a masculine technical culture by discussing more "concrete" and "bricolage"-focused methods of learning to program, which they claimed that women and girls prefer.[99] Likewise, many members of the broader constructionism community including Cynthia Solomon, Sylvia Weir, Yasmin Kafai, Ricarose Roque, and more have focused specifically on how constructionism might better appeal to girls and other diverse groups.

Still, Papert constantly referred to the cultural trappings of technically precocious boys as integral aspects of his constructionist dreams: gear sets and other engineering-oriented toys to tinker with, computers to program, and video games to play, all of which have a long and deep association with boys' play, boys' culture, boys' cultural identities, and boy-focused social imaginaries. This signals that boys are a natural fit while girls are marked as exceptional and will thus need to account for themselves—much like women continue to have to do online and in technical spaces. Thus, although the development of constructionism more broadly owes a great deal to the many women who have been part of constructionist projects over the years, Papert's descriptions of it still lean heavily on ideals from MIT's overwhelmingly male hacker culture. One Laptop per Child—largely populated by men who also identified with this hacker culture—did little to challenge these masculine defaults.

Social Imaginaries and Identity: The Self-Taught Hacker

The social imaginary of the technically precocious boy, who rebels against factory-model schooling and authority more generally, helped establish the charisma of Papert's constructionist projects, from Logo to One Laptop per Child.[100] This social imaginary reinforced certain group identities and values within OLPC and across the technology world more generally, reflecting the way that "childhood" tends to be imagined and articulated through the design process in many academic and industrial settings.

These actors also referenced these social imaginaries of childhood when discussing *themselves*—and it is through this pathway that charisma becomes personal, with implications for technology design that we will explore in the next chapter. In his writing, Papert described himself and other hackers like him as part of a rarefied group that had managed to hold on to the magic of childhood: they were playful, rebellious, independent thinkers and lifelong learners. In Papert's *yearners*, we can likewise recognize the MIT hackers' conception of themselves: adults who held on to the subversive, insatiable curiosity that Papert believed all children have but often lose through traditional education. In their own view, MIT's hackers had resisted the confining, stifling effects of instructionism and—with the help of computers—remained passionate learners and free thinkers. After all, Papert explicitly credited his own experiences with this hacker group as inspiration for constructionism and as the reason for advocating for one computer per child.

These imaginaries thus not only implicate group membership, they can, by extension, shape one's sense of identity. Here, the social imaginary of the technically precocious boy provides a particular archetype around which people can shape their own identities, whether in agreement or opposition. The pervasiveness of these social imaginaries means that we all recognize them, but our reactions to them can differ. We each negotiate our own unique "subject position" in relation to them, as cultural theorist Stuart Hall would call it.[101] OLPC's contributors appeared to take a hegemonic approach to the social imaginary of the technically precocious boy by assuming that it would be universally recognized and universally accepted because they saw it as the natural state of children, not as a socially constructed imaginary at all. Such naturalization makes these imaginaries—and the projects, careers, and identities built on them—particularly thorny to unravel. Papert, for one, may have been so resistant to revising his constructionist theories even when evidence failed to bear them out because to do so would have entailed giving up some of the values that appeared to have been foundational to much of his professional, if not his personal, identity.

Moreover, the way that the social imaginary of the naturally creative yearner frames individualized creativity as innate shores up the American cultural values of inborn greatness and rugged individualism, and also erases the extent to which cultural expectations have differently shaped the play of different groups of children within and beyond the United States.[102] Treating childhood as a natural state that is beyond the trappings and obligations of culture means that children's toys—OLPC's laptop included—can then be exported around the world without concerns of cultural imperialism. In this way, those who build such devices around imaginaries based deeply and inextricably in US cultural norms—and, in particular, norms that define white, middle-class American boyhood—can export them to children in the Global South without feeling that they need to account for what different contexts those devices might be entering. Instead, these toys tap into the placelessness and timelessness that the imaginary of the naturally creative child evokes among those who share it.

Overall, alignments between constructionism and the ethos of the hacker community at MIT, of which Papert was a part, have meant that even in the face of contradictory evidence, Papert's constructionist projects (such as Logo) and texts (such as *Mindstorms*) have remained charismatic for many in the technology world. When OLPC was built around constructionist principles, the project adopted—and amplified—the charisma from which Logo and *Mindstorms* had also benefited. Moreover, over fifty years after Papert introduced his ideas and over thirty years after educational researchers demonstrated that those ideas did not work as advertised, variations of constructionist learning are still being celebrated at technical conferences and taught in technology-design schools around the country.[103] Not only are Papert's former students and others who experienced the Papert-heavy MIT Media Lab curriculum employed in many of those schools, but the hacker ethos that constructionism was built on—an ethos that draws on some of the foundational beliefs that American culture has about children, rebellion, and technical creativity—still influences the technology world more generally. The next chapter will explore how these imaginaries played out in the design of OLPC's laptop—a process I call "nostalgic design"—and the risk they faced of reinforcing existing inequities through the sexism inherent in these imaginaries. Exposing the cultural roots of the imaginaries, as we have done here, allows us to critically examine the relationship between the designer, the artifact, and the kind of user they hope to reach.

2 Making the Charisma Machine: Nostalgic Design and OLPC's XO Laptop

> This is in some ways a Trojan horse. Governments are buying into this because this looks like a terrific way of creating a computer-savvy workforce, fighting brain drain, essentially bursting into the twenty-first century. But what it also is is a backdoor into overhauling the entire education system of a lot of the countries that we are talking about.
>
> —Walter Bender, quoted in Lydon, "One Laptop per Child?"

While Nicholas Negroponte traveled the world in 2006 giving talks about the hundred-dollar laptop project, the newly formed nonprofit One Laptop per Child got to work designing the laptop itself. The foundation incorporated in Delaware in 2005 and in March 2006 moved out of the MIT Media Lab and into an office in Cambridge, Massachusetts, about a mile away.[1] There, a core group of around twenty people—many affiliated with MIT but some coming from as far as England and Italy to work on the project— did the bulk of the work designing the laptop hardware, developing the software, and coordinating with potential client governments and donor companies throughout 2005, 2006, and 2007.[2] A much-larger group of developers—hundreds, based on OLPC mailing-list membership records— worked part time on OLPC's open-source software projects, coordinated by this full-time staff.

Based on their lively discussions in public talks and on public mailing lists and on my later interviews with some of its members, this group was animated by the ethos we explored in the previous chapter—the MIT hacker community's love of programming, mistrust of authority, and belief in computers as a force for positive social change—as well as the social imaginaries that were foundational to this ethos, particularly the innate creativity and rebellious nature of the archetypal technically precocious boy and its counterpoint, the stifling school as factory. We saw that they were likewise united by their conviction that MIT Media Lab professor Seymour Papert's theories of constructionist learning with computers could inspire these kinds of children to teach themselves how to, as Papert put it, "think about thinking" and to "learn how to learn," building from programming skills to become autodidacts more generally.

These convictions provided the foundations for the project's charisma. Its promise of social transformation that centered on many of the values they held most dear gave its members a strong sense of purpose. This charisma was not focused on the material form of the device itself, though this chapter will show how these ideals influenced OLPC's design of the laptop. Rather, it is focused on what effects these features were meant to have in the world; what mattered was not the object's materiality so much as what it promised to do. As OLPC's mission originally asserted, "OLPC is not at heart a technology program and the XO [laptop] is not a product in any conventional sense of the word. We are non-profit: constructionism is our goal; [the] XO is our means of getting there."[3]

Moreover, OLPC's laptop was not only meant to change the world but to be a shortcut to social change, succeeding where other programs had failed. In a 2007 interview about OLPC at the MIT Museum, founding project member and MIT Media Lab senior research scientist Walter Bender called the laptop project an "end run" to significant change: "We are certainly supportive of efforts to build more schools, get more teachers into those schools, but we think that that's sort of a treading-water process, and we're trying to do something that can do a little bit of end run around the status quo and move a little bit quicker to reaching more children, and we think that the connected laptop is a part of that solution."[4] In publicly available mailing lists, blog posts, meeting transcripts, and talks, OLPC employees on the project expressed similar sentiments. In a 2007 talk at Google, developer Ivan Krstić said, "We are way too impatient, we being OLPC. ... Can't we make this better, right away?" He concluded, "Let's get learning driven by curiosity again."[5] OLPC employee and Red Hat programmer Chris Blizzard stated at a 2007 analyst meeting that OLPC would have an impact on education, social issues, and "almost anything" by "empowering a population to look at the world around them to have the initiative."[6] Negroponte stated this even more forcefully. In a 2006 interview, he said that what inspired everyone at OLPC was eliminating poverty through the education their laptop would provide. "That's why everybody who's involved in the project is involved with it," he said. "You can eliminate poverty with education, and no matter what solutions you have in this world for big problems like peace or the environment, they all involve education."[7]

Clearly these employees were taken with the laptop's charisma and shared a vision of what this charisma would enable in the world: harnessing an innate curiosity about technical systems, children across the Global South would joyfully take up OLPC's computers and, in the process of teaching themselves, lift themselves, their families, and their societies out of poverty. This chapter examines how these charismatic stories of easy, child-focused, laptop-driven social change influenced the design of the laptop that OLPC built. What factors and features were considered most important, why, and what do they tell us about the organization's vision and priorities? In particular, how did the social imaginary of the technically precocious boy (which we explored in the previous chapter) play into this, and how much did contributors lean on rose-tinted memories of their own first encounters with computers in childhood, a process I call "nostalgic design"? In short, how was the machine designed to be charismatic, and how was its charisma expected to act on the world?

Charismatic Potential, Nostalgic Implementation

As a place to begin, we can consider excerpts from the project's five core principles, first published on One Laptop per Child's website in 2006 and often reiterated in presentations and discussions about the project:

Principle 1: Child Ownership. "The ownership of the XO is a basic right of the child and is coupled with new duties and responsibilities, such as protecting, caring for, and sharing this valuable equipment. ... A key OLPC asset is the free use of the laptop at home, where the child (and the family) will increase significantly the time of practice normally available at the standard computer lab in the school."

Principle 2: Low Ages. "The XO is designed for the use of children of ages 6 to 12—covering the years of the elementary school—but nothing precludes its use earlier or later in life. Children don't need to write or read in order to play with the XO and we know that playing is the basis of human learning."

Principle 3: Saturation. "We need to reach a 'digital saturation' in a given population ... where every child will own a laptop. ... A healthy education is a vaccination, it reaches everybody and protects from ignorance and intolerance."

Principle 4: Connection. "The XO has been designed to provide the most engaging wireless network available. The laptops are connected to each other, even when they are off. If one laptop is connected to the Internet, the others will follow to the web. ... The connectivity will be as ubiquitous as the formal or informal learning environment permits. We are proposing a new kind of school, an 'expanded school' which grows well beyond the walls of the classroom."

Principle 5: Free and Open Source. "The very global nature of OLPC demands that growth be driven locally, in large part by the children themselves. ... In our context of learning where knowledge must be appropriated in order to be used, it is most appropriate for knowledge to be free. Further, every child has something to contribute; we need a free and open framework that supports and encourages the very basic human need to express. Give me a free and open environment and I will learn and teach with joy."[8]

The first two principles articulate the project's intended beneficiaries: elementary-school-aged children. Whereas the first focuses on the responsibility of owning a computer, the second turns to the play that the computer is meant to enable—an important theme that motivated many of the features of the laptop. The third principle specifies that to have transformative effects, this laptop should be given to not just some children but to all. The fourth centers on the promise of connectivity—to one another through the laptop and to the internet—which along with play shaped the design of the laptop. And the fifth, focused on the freedom that open-source software was meant to provide, most closely captures the commitments of MIT's hacker culture and the open-source software movements that grew from it. This principle was a crucial element of the project's charisma for many of its contributors. Three key themes present in these principles—*play*, *connectivity*, and *freedom*—constitute the key pillars of OLPC's charisma, and all were built into its laptop. We will explore each theme in turn below.

Also threaded throughout all three of these themes, as we will see, is a sense of nostalgia for the kinds of computing experiences that many contributors recounted about their formative years (childhoods for some, early adulthoods for others). Nostalgia, a sentimental attachment to an idealized or imagined past, can be a potent social force. Stephanie Coontz explores how collective nostalgia for a "typical American family" that has always been more myth than reality has resulted in strong social norms and even policies across the United States in attempts to enact it.[9] Svetlana Boym explores the role of nostalgia among diasporic Russians and concludes that nostalgia "has a utopian dimension" as both "a sentiment of loss and displacement, but ... also a romance with one's own fantasy"—a romance that ends if faced with reality.[10] Although nostalgia is often invoked dismissively, my intent here, like Boym's, is to engage with what makes it powerful and how it shapes not only social worlds but the design of charismatic technologies.

In One Laptop per Child, for instance, nostalgic stories often emerged in discussions between OLPC employees and within the larger OLPC community across the internet. Moreover, in addition to a desire to minimize power usage, we will see that nostalgia also influenced the decisions the team made in designing OLPC's laptop, including the choice of hardware components for the machine, which went into mass production at the end of 2007 and was incrementally updated in 2009. Indeed, the laptop's hardware was otherwise puzzling. The first-generation laptop's 433-megahertz processor was nearly ten times slower than state-of-the-art machines in the mid-2000s; it was typical of entry-level machines in 1999.[11] It had a one-gigabyte storage capacity on solid-state Flash memory, which employees said would be faster and lacked the jammable moving parts of the then-common hard disk drive. However, this storage capacity was miniscule for the burgeoning multimedia-rich computer landscape: whereas one gigabyte was standard for a computer from the mid-1990s, computers in the mid-2000s typically came with fifty- to one-hundred-gigabyte hard drives. Likewise, its 128 megabytes of RAM (random-access memory, a computer's working memory) was one-quarter of the 512 megabytes typical of entry-level machines in 2005. Slow as they were, each of these specifications was reportedly at least double what the team had first prototyped.[12]

A machine with hardware that was a decade behind current entry-level devices was, in a way, the point. Project leaders, including Bender and Jim Gettys, said it meant that the laptop consumed less power and its components were cheaper to source, and my discussions with OLPC contributors confirm that these were the primary motivations.[13] The "OLPC Myths" page of the project's official wiki admitted that the laptop "isn't powerful enough to run modern 3D games and other resource-heavy programs such as video-editing software." But, it continued, this was "irrelevant: that's not the purpose of this laptop. It is designed to be an inexpensive way for people of limited means to use a computer for such things as internet and educational software. The choice is not currently between this system and a more capable one: it is between this and nothing. This is better."[14]

Such a stance, however, opened the project up to critics who accused it of thinking that children in the Global South did not know or deserve better (and as we will explore in subsequent chapters, these hardware choices did introduce problems in use). So how did contributors square the creation of such a deliberately underpowered machine with the revolutionary potential they saw in it? In interviews and on public mailing lists and blogs, some justified it by comparing OLPC's laptop not to contemporary machines but to the computers they had used in their youth, such as the Commodore Amiga and Apple II computers on which they had learned to program, or the Atari and Commodore 64 game systems they had hacked. These computers, after all, had much less power, but these users had still found them captivating.

Moreover, like old cars, these vintage computers had been simple enough that they could be understood more completely, from the hardware up to the interface. On public OLPC mailing lists and other discussion forums online, some OLPC community members lauded OLPC's laptop as a return to the good old days of computer hacking, when every memory location could be tracked and every pin on a connector controlled; they said that contemporary systems, in comparison, were too hopelessly complex and layered to enable that kind of low-level hacking. Emails to OLPC mailing lists and discussions on the project's wiki compared the specifications of their laptop with Amiga systems and other early computers. A fan who received a "give one, get one" laptop in early 2008 even ported the Amiga operating system to an OLPC laptop—a feat that gained him accolades across the OLPC community and wider internet.[15] Other classic programs and video games were ported as well.

This chapter explores how this justification for these hardware choices is an example of *nostalgic design*: it targeted the social imaginaries that resonated best with OLPC contributors' own identities, especially the social imaginary of the technically precocious boy who found joy in understanding the machine deeply. However, nostalgic design violates some of the core principles of user-centered design: to design for one's users, not for oneself, and to account for the messy realities of use.[16] Indeed, the project did little to engage these realities: there was no OLPC-specific pilot or user testing to speak of.[17] Negroponte, in fact, derided pilots in his talks: in his 2006 TED talk on OLPC, he said, "The days of pilot projects are over. When people say, 'Well, we'd like to do three or four thousand in our country to see how it works.' Screw you. Go to the back of the line and someone else will do it, and then when you figure out that this works, you can join as well."[18] As reported by MIT professor Ethan Zuckerman in a 2006 update on the project, Gettys told him that "the current plan to distribute five million laptops in five nations next year *is* a pilot—when you're talking about building and distributing more than two billion devices, a few million is just a toe dipped into the water."[19]

Though this flies in the face of well-established usability principles, designing for some abstracted user based more on social imaginaries or one's own nostalgic memories instead of taking the time to really grapple with the contradictions of actual use is nevertheless common in technology design. It crops up even in practices meant to be user centered, such as scenario building and persona creation, in which especially less experienced designers can easily lean on stereotypes in lieu of actual user contact. It can be applied without really intending to do so, because it is easy to fall back on social imaginaries and idealized users—and especially on one's own rosily remembered past—rather than grapple with the complications of real use by real people who might be quite different. Genevieve Bell and Paul Dourish have shown how past visions of the future as depicted in science fiction continue to animate the field of ubiquitous computing, for one.[20] Similarly, Fred Turner has linked the initial vision of "cyberspace" portrayed in William Gibson's *Neuromancer* as pivotal in defining the internet as an "electronic frontier."[21]

A form of nostalgic design is more consciously harnessed by the MIT Media Lab, OLPC's birthplace. One of the school's core graduate courses, which has been taught for several decades, is built around using students' own childhood experiences as inspirations for design. This is guided by Papert's preface to *Mindstorms* where he discusses how the "gears of his childhood" inspired his development of constructionism. Papert, for one, openly admits that his own childhood provided both inspiration for constructionism and fodder for his descriptions. "I shall in fact concentrate on those ways of thinking that I know best," Papert explains in the first chapter of *Mindstorms*. "I begin by looking at what I know about my own development."[22]

Thus, nostalgic design's reliance on collectively held social imaginaries as design inspiration was crucial for building the charisma of OLPC and for sustaining the motivation of the project team and the larger community of contributors. Through this, they were able to maintain a shared vision for what the project would accomplish in spite of what looked like insurmountable odds—and to shore up this vision against many of the compromises they had to make along the way.[23] In the balance of this chapter, we will explore how nostalgic design influenced the XO laptop's key features, which are grouped thematically into play, connectivity, and freedom. We will see that *play* is reflected in the toy-like appearance of the laptop as well as in the proliferation of games, game engines, and game-oriented constructionist software on the device—a proliferation that reflects the importance of video games in the childhoods of many of OLPC's developers. *Connectivity* was built into the XO via its "mesh network," which was designed to enable easy connections to other laptops and the internet and taken as a prerequisite for its Bitfrost security system—and which reflected the kinds of computer-mediated connections that were most important to many on the OLPC team. Encoding *freedom*, the XO laptop's software was all free and open source, and the XO keyboard had a View Source button to emphasize this. The team also tried to make the laptop hardware open source and easy to repair—efforts that reflected their conviction that children can and should teach themselves to program, just as they remembered teaching themselves. Beneath these themes was an unspoken faith in the XO laptop's ability not only to succeed in its constructionist goals but to be a shortcut to solving social problems where other approaches had failed—a faith in technology as a source of radical social change.

The Promise of Play

The longer description of One Laptop per Child's second core principle, "Low Ages," asserts that "children don't need to write or read in order to play with the XO and we know that playing is the basis of human learning."[24] What kind of play are we talking about? In the previous chapter, we saw that Papert centered his investigations of constructionism on open-ended play on a computer and asserted that technical exploration was an important kind of play. We also saw that the first-generation MIT hackers whom Papert spent time with as part of Marvin Minsky's computer lab focused on playful exploration of computers as well. They created some of the first computer games, such as the now classic game Spacewar!, and spent many hours playing one another's creations.[25] Papert was so taken with this community that he described his experiences with them as "playing like a child" and "pure play."[26]

These kinds of nostalgic memories of computer-centered play shaped the choices that OLPC made about the physical design of the XO laptop. Its cute appearance made it look like a toy (which we will see was a double-edged sword in use), designed specifically for children. The laptop's bright colors and curved lines evoked the Apple iMac and iBook computers of the late 1990s and early 2000s, which had been designed as a playful break from the black and beige norms of computer hardware at the time. After OLPC explored all green and all orange as possibilities, they settled on a textured white laptop outlined in bright green rubber and with a scaled-down green, silicone-membrane keyboard made for children's smaller fingers.[27] Its 7.5-inch screen swiveled 90 degrees in one direction and 180 in the other, where it could be lowered into a low-power e-book mode. Replacing the iconic but infeasible hand crank from the November 2005 debut demonstration were two green "ears" (Wi-Fi antennae), which together covered one audio port and three USB ports and latched the laptop shut.[28]

The project's logo, created by Michael Gericke of the design firm Pentagram, was a rounded, sans-serif X with a dot nestled above it, meant to evoke a child with arms and legs flung wide in the excitement of play.[29] This logo also matched the laptop's eventual name: the XO. During its development in 2005 and 2006, Negroponte and others on the project variously called OLPC's first-generation laptop the "green machine," "children's machine 1" (or CM1), and "2B1."[30] In fall 2006, the group settled on "XO"—the logo on its side. This took inspiration from the text-based emoticons that had originated in the early 1980s, such as : -) . One large XO logo, in various colors, covered the top of each laptop, and tiny XO logos made up its bumpy texture.[31]

But even more central to the project was the team's strong commitment to including playful software on the laptop. Unique in the ecosystem of classroom laptops, the XO included not only custom-designed hardware but a whole software suite geared toward play. Its graphical user interface—designed by Pentagram and called "Sugar," a cheekily intentional reference to something children love but adults are not so sure about—tried to introduce to children a new paradigm of computer use centered on playful metaphors. Gone was the standard windowing system and desktop, with its office connotations. In its place was a playground of things to do. The child's XO logo was in the center and programs—called "activities"—were arranged around it.[32] Clicking on an activity brought it up full screen, generally with a menu of commands along the top. Children could click on the stop sign in the upper right corner to close the program and return to the playground. Aside from this playground, there were two other views: the friends view, which showed other computers that had been connected in the past, and the neighborhood view, which showed all nearby Wi-Fi transmitting devices detected by the laptop.

This Sugar environment moreover shipped with several activities that had been specifically built around the idea of constructionism at the MIT Media Lab and elsewhere. It naturally included Papert's signature constructionist project developed in the 1970s and 1980s: the Logo programming language and its iconic turtle, packaged for the XO as Turtle Art. Also included was the programming environment Scratch, which became widely popular through the 2010s but was still up and coming in the mid-2000s, spearheaded by Mitch Resnick, an MIT Media Lab professor and Papert's former student. Scratch's design lent itself to storytelling and game creation, and a website to encourage sharing and "remixing" Scratch code—though initially developed as a side project by Media Lab student Andrés Monroy-Hernández in 2008—developed into a large online community of open-source Scratch projects. The central mechanism for interaction in both Scratch and Turtle Art was block-based programming: programming commands were encapsulated in drag-and-drop shapes that only fit together if they were compatible, and advanced students could create their own programming blocks. In keeping with the principles of constructionism, these programs all framed play as something involving programming on a computer.

Another related aspect of play also emerged from the compilation and discussion of the laptop's software: playing video games. Citing the importance of video games in their own childhoods, community members quickly galvanized around making games and game engines available on the laptop. Though each country could choose a customized "activity pack" of software to include on the laptop, most used a fairly standard package that included several video games.[33] Additional activities were available for download from OLPC's wiki and from third-party sites.[33] They included dozens of games of all sorts, from familiar video, card, and board games that were ported to the XO (including Spacewar!) to games custom designed by the open-source community. During the first-generation XO laptop's development in 2006 and 2007, the Games mailing list was among the most active of over seventy public lists that OLPC hosted.

On this list and elsewhere, various OLPC developers justified the inclusion of games on the XO laptop and the proliferation of game options by citing constructionism's focus on play. The OLPC wiki page that describes the games available on the XO says, "The OLPC is an education project but that doesn't mean there is no room for fun. On the contrary, fun must be integrated into the child's educational experience."[35] But this conflation of open-ended play and video games exposed some of the ideological commitments of the project, especially within the context of broader cultural debates about the value of video games.[36] It also ignored the ways that video games and gameplay have been gendered by game companies and other media since the 1980s, as we saw in the previous chapter. If play and video games were so closely linked, then it was boys that OLPC was implicitly targeting in its focus on play.

Of course, different games elicited different arguments. When Chris Blizzard ported the first-person shooter game Doom—groundbreaking in 1993 for its 3-D graphics—to the first-generation XO, its existence and ease of installation caused controversy even among the game enthusiasts on the Games mailing list due to its violent content.[37] However, some, such as open-source programmer and OLPC software contributor Albert Cahalan, argued that even violent first-person shooter games such as Doom are educational and "have led many people to study things like geometry and physics."[38]

Similarly, both Negroponte and Papert have written that all games offer some educational benefit in teaching strategic thinking, the value of cooperation and competition, and a passion for learning. When making the case for the educational value of games in *Being Digital*, Negroponte directly addresses a presumably video-game-playing, or at least video-game-familiar, audience that is motivated by competitive learning: "Most adults fail to see how children learn with electronic games. The common assumption is that these mesmerizing toys turn kids into twitchy addicts and have even fewer redeeming features than the boob tube [television]. But there is no question that many electronic games teach kids strategies and demand planning skills that they will use later in life. When you were a child, how often did you discuss strategy or rush off to learn something faster than anybody else?"[39] Likewise, for those who didn't think video games had anything educational to offer, Papert counters with a description of the complexity of "most" games, collapsing this wide-ranging and heterogeneous genre into one characterized by "complex information" and "fast-paced" play:

Any adult who thinks these games are easy need only sit down and try to master one. Most are hard, with complex information—as well as techniques—to be mastered ... These toys, by empowering children to test out ideas about working within pre-fixed rules and structures in a way few other toys are capable of doing, have proved capable of teaching students about the possibilities and drawbacks of a newly presented system in ways many adults would envy. Video games teach children what computers are beginning to teach adults—that some forms of learning are fast-paced, immensely compelling, and rewarding.[40]

The developers' plans for games on the XO further paralleled the gaming of MIT's hacker community by including not just games but also game engines, or programming environments and code libraries that are meant to make the creation and modification of computer games easy.[41] Just as MIT's early hacker community was driven to learn clever programming tricks to write and improve computer games, OLPC developers hoped that children would be motivated to learn how to program computers so they could, among other things, write and improve games on the laptop. In addition to Turtle Art and Scratch, which could both be used as game engines, the standard XO activity bundle included eToys, designed to enable game creation. Other game engine options included Frotz, ISIS, Pygame, Boardwalk, and Physics. And there was a push for even more. In the description of the first OLPC Game Jam in Boston in 2007—where OLPC engaged the broader open-source community in programming games for the first-generation XO over a long weekend—OLPC employee S. J. Klein wrote, "More importantly it is hoped that kids can use the laptops to create their own games and experiment deeply with learning games by having access to modify and change them as part of a learning process."[42]

Why were developers so sure that kids around the world wanted to make games? Though video games are widely popular, most children who play video games do not go on to make them, and not all children who have access to these kinds of tools decide that is how they would like to spend their time. OLPC seemed to implicitly expect that children using the XO laptop would adopt their definition of play as one that involved video games and competition and one that ultimately led to programming—a starting point to learning deeply about computers and adopting the values of the programming community. In this way, they projected their own values onto the children they hoped to reach. They assumed that these children would find the same kinds of play captivating, and that this play would lead to the same positive outcomes in these children's lives as it appeared to have in their own.

The Promise of Connectivity

OLPC's fourth core principle, "Connection," refers to one of the project's most hotly anticipated features (perhaps after its nonexistent hand crank): its mesh network. Though the mesh network did not work in practice and was in fact dropped in an update to OLPC's Sugar software during my field-work in 2010, it generated much excitement in the programming community and technical press when it was announced. It would let XO laptops connect to one another without the need for an access point to moderate the connection, allowing them to circumvent any surveillance, censorship, or other attempts at control. It would let a group of laptops share one internet connection that might be on one edge of town (never mind how abysmally slow that connection might become). Even when it was off, a laptop could act as a relay in the network (never mind the battery-draining implications of this). The software for the XO laptop was designed to take advantage of this network and be used collaboratively. In theory, XO users could share their screens with others, collaboratively create content (from music to text to drawings to code), and automatically propagate connections to the internet, all at the click of a button. In anticipation of this feature, the XO software also included visualizations of other laptops nearby: the neighborhood view and friends view.[43]

Why this focus on connecting computers? In constructionism, collaboration with other students is meant to enable peer learning and help students better learn from their mistakes (or "debug," a programmer's term that constructionism appropriated to describe this). Papert observed this peer learning occurring when one child looked over another's shoulders or worked together on the same machine, but the XO laptop focused on computer-to-computer connections where each child was at their own machine, across the classroom or even the town.

This focus on connection was also a result of nostalgic design, reflecting developers' own memories of positive experiences online in their youth. Online spaces such as bulletin board systems (BBSs), chat rooms, Internet Relay Chat (IRC) channels, blogs, multiplayer online games, and social media are not places of harassment and trolls, they argued; they are places where computer-loving iconoclasts, technically precocious boys like many of them had been, could really be themselves, find others like them, and broaden their horizons. Steven Levy captures this sentiment well in describing MIT's hacker culture: "These young adults who were once outcasts found the computer a fantastic equalizer."[44] If computers are an equalizer in the eyes of people who share these values, then increased connectivity and communication via computers is naturally a positive force in children's lives. A belief in the positive power of this connectivity is reflected in the extended descriptions of OLPC's third and fourth core principles. "Because of the connectivity inherent to OLPC these different communities will grow together and expand in many directions, in time and space. They will become solid and robust, because they are saturated, without holes or partitions," read the extended description of principle three, Saturation. Principle four, Connection, further elaborates, "This connectivity ensures a dialogue among generations, nations and cultures. Every language will be spoken in the OLPC network."[45]

This connectivity was also meant to help children subvert authority figures, reflecting the disdain for authority and for school that was present in both constructionism and the hacker community. Constructionism instead casts the teacher as a "co-learner" but then says children in a constructionist environment may soon outstrip teachers in knowledge.[46] In the previous chapter, we saw that Papert used a monolithic caricature of school as a foil for playful learning, relying on the social imaginary of school as factory; to him, school was an institution that was designed to crush children's natural creativity. We saw that others in OLPC's leadership likewise ridiculed teachers, particularly teachers in the rural Global South, whom they accused of being undereducated, undertrained, "drunk," or even "absent entirely."[47] Ivan Krstić said in a talk at Google about OLPC, "Kids should be able to have some kind of way to get curious and get answers regardless of whether their teacher can provide them or doesn't know or isn't there."[48] In an early description on his blog of the project's progress, Ethan Zuckerman further explains how the team was then thinking about teachers:

> No matter how you slice it, the laptop is a deeply subversive creature, likely to undercut the authority of teachers who don't figure out how to master the device as quickly as their students. ... It's too easy for me to imagine teachers threatened by the laptop ordering students to put them away and watch the blackboard.

Walter [Bender] and crew aren't unaware of these issues. He points out that the machine is a laptop precisely so students can take it home and learn with it in spite of their teachers.[49]

OLPC has since softened this harsh opinion of teachers, especially as teachers have turned out to be crucial in motivating children's use of the XO laptops (just as they played a pivotal role in the spread of Logo in the 1980s, despite being caricatured in similar ways in Papert's early writing). However, the laptop's mesh network and collaborative software—and its lack of any interface for teachers—still reflected these attitudes. In 2007, the project added a "school server" (or XS), but this was designed to create backups, store content locally, renew XO security licenses, and push software updates; it still had no tools to help teachers.[50]

An extension of the need for connectivity, but a counterpoint to this antiauthority ethos, existed in Bitfrost, OLPC's security system. This system was designed for the XO laptop by Ivan Krstić, OLPC's director of security architecture (and later head of security engineering and architecture for Apple Computers). It would shut down any XO laptop that had not "checked in" to a school in a specified period of time (by default one month) and would permanently disable any that had been reported as stolen. Here, the team's desire for decentralized connectivity was overcome by its paranoia that these laptops would be stolen by strangers or even confiscated by parents and sold on the black market.[51] So, in the name of computer security (and with the assumption that connectivity, at least to the school server, would not be an issue), Krstić created a BIOS-level security lock that could not be deactivated and also managed programs' access to hardware peripherals, such as the camera and microphone—all without the need for a password on the machine.

The Promise of Freedom

The last, and in many ways the most central, theme that directed the design of the XO laptop was the goal to make it as open source as possible. OLPC's commitment to using free and open-source software (and, where possible, hardware) was stated explicitly in the project's fifth principle, "Free and Open Source," and was reiterated by project employees. In an interview on National Public Radio's *Radio Open Source* program in February 2007—in which the XO laptop was described by the commentator as "a hacker's paradise, a real open-source playing field"—Bender elaborated that an open-source laptop allows for deep exploration: "We want the children to be able to reach inside the machine, ... and a closed system does not allow that."[52] Chris Blizzard articulated the connection to free and open-source software more directly by stating that it is necessary for learning: "The commitment to open source and free software is still one of the main [principles] of the project. Learning requires the transparency that free and open source software provides."[53] The connection between this commitment and the ethos of MIT's hacker community is straightforward; indeed, a number of this community's members, such as Free Software Foundation founder Richard Stallman, built their careers and lives around it. This group's passion for free and open-source software is central enough that it appears in the hacker ethic as Steven Levy articulated it: "all information should be free" and access to computers should be "unlimited and total."[54]

The primary and easiest target of this commitment was the laptop's interface, Sugar, which was built on top of the Red Hat version of the (open-source) Linux operating system and was all open source itself. In the spirit of the software logs on Linux, moreover, opening an activity also made an entry for it in a Journal program, which was meant to be a record of all of the learning a child had accomplished with the machine—though in practice it was a memory hog (as Logo also was in the 1980s) that quickly filled with junk entries that users had to manually clear.[55]

But the goals for open source went beyond the laptop's software to include its hardware. The team wanted to make the laptop's hardware easily "hackable" by children, which would both help with repairing laptops and with modifying them in unexpected ways. "The [mother]board itself is designed to encourage hardware hacking," Zuckerman explained in the 2006 project update on his blog. "Want to turn a laptop into a device that can drive an external monitor? Solder one [a video graphics array (VGA) jack] on."[56] OLPC hoped that its hardware design would encourage kids to tinker with the laptop and even to be able to repair it themselves, which would let them really understand the machine inside out, the way the developers themselves understood computers. Like the old hardware specification described above, this illustrates another clear instance of nostalgic design: it assumed a world with easy-to-obtain peripherals that were compatible with one another, such as the self-designed and easily repairable desktop "tower" computers that were especially prevalent in the 1990s.

Though critics called the expectation for children to repair their own laptops naïve or exploitative, Negroponte, Papert, and others defended it as the perfect learning experience, providing an outlet for children's supposedly natural proficiency with technology and allowing them to delve as deeply as possible into the workings of the machine. In a USINFO "Webchat" interview about OLPC in 2006, Papert claimed that having children repair the laptops was a matter of empowerment, not exploitation: "I believe in 'Kid Power.' Our education systems underestimate kids. It INFANTILIZES them by assuming they are incompetent. An eight-year old is capable of doing 90% of tech support and a 12 year old 100%. And this is not exploiting the children: it is giving them a powerful learning experience."[57]

Some of their design goals were realized. The first-generation XO-1 has no moving parts such as fans or a spinning hard drive to jam or break when dropped. It has minimal connectors to get broken or dirty—just one Secure Digital (SD) slot, three USB slots, and one audio jack, all covered by the laptop's hinge or antennae when the laptop is latched, plus a power connector. It has a solid, silicone-membrane keyboard to make the laptop water resistant, a rugged case to protect it from falls or other wear and tear, and standardized screws—with extras included inside the handle—to make dismantling and repairing the laptop easier.[58] But a former OLPC employee informed me that the goal of having the XO's various components soldered onto the machine's motherboard rather than permanently glued, to make it possible to swap them out, was scrapped by OLPC's hardware manufacturer Quanta, which had different priorities and constraints—especially keeping the laptops low in cost and its vendors happy.

Though it did not realize its goal of completely hackable laptop hardware, the team did win one concession. It added a dedicated key on the keyboard, to the left to the Space Bar, that was meant to show the source code for any activity on the laptop. Bridging hardware and software, this View Source button—symbolized by a gear, Papert's favorite childhood object-to-think-with—joined the mesh network as one of the laptop's most discussed features. OLPC developers hoped that this would enable children to learn how to program by being able to make clear connections between source code and what a program does, much as the View Source option in web browsers have helped people learn HTML (a comparison often made in online discussions about the feature). It would, in theory, also encourage users to change features of XO activities to better match what they want.[59]

As Bender explained, "With just one keystroke, the Sugar 'view source' feature allows the user to look at any program they are running and modify it. The premise is that taking something apart and reassembling it in different ways is a key to understanding it."[60] Even though I never saw the View Source button used or even acknowledged by children or teachers throughout my fieldwork, it became a particularly evocative feature among developers and others tracking the OLPC project because it embodied programming culture's values so well. After OLPC announced the feature, in fact, others proposed adding View Source buttons to other programs and systems.

These commitments to free and open-source software embodied many of the tenets of Papert's constructionism, namely its interest in programmable computers as objects-to-think-with, objects that have "no ceiling" and can keep growing with learners throughout their lives. They also connected with many of the core values of hacker culture at MIT and beyond. With the XO, Papert's ideas for a constructionism-based "children's machine," which he first articulated in the 1960s, were finally realized. Benjamin "Mako" Hill, who was then an advisor to OLPC and a board member at the Free Software Foundation during his graduate work at MIT, articulated the connections between constructionism and open-source software more explicitly: "Constructionist principles bear no small similarity to free software principles. Indeed, OLPC's stated commitment to free software did not happen by accident. OLPC convincingly argued that a free system was essential for creating a learning environment that could be used, tweaked, reinvented, and reapplied by its young users." Only on a completely open system, Hill explained, could children really deeply explore, with no limits to what they could discover. Together, they "offer a profound potential for exploration, creation, and learning. ... Free software and constructionism put learners in charge of their educational environment in the most explicit and important way possible. They create a culture of empowerment."[61]

The XO laptop's open-source software served an additional function: it enabled OLPC contributors to make a case that the laptop was value neutral. In the same blog post, Hill also explained this line of reasoning: "We know that laptop recipients will benefit from being able to fix, improve, and translate the software on their laptops into their own languages and contexts. ... If you don't like something, change it. If something doesn't work right, fix it."[62] In another blog post, he pushed back against critiques of OLPC as culturally imperialist by framing its open-source software and programmable, open-source devices as a means of avoiding cultural imperialism:

> People in the rich and developed countries may have cellphones but they frequently also have computers: full-fledged, reprogrammable, hackable computers; computers that they can use to write software, design hardware, install new OSes [operating systems] on, and even—if they are really adventurous—use to reprogram their mobile phone.
>
> People in the developing world will have information technology (in the form of cellphones at least) but do not have the ability—no matter how interested, talented, or intelligent they are—to change the way they work. *This* is the greater danger. ...
>
> In three years, there will be a billion people in the developing world who are using information technology on the terms and at the whim of today's global elite and they will *not* be able transcend their role as consumers and subservients [*sic*] in this context. ...
>
> Unless we do something about it. ...
>
> That is my personal goal in OLPC and it is one that has seemed to have been echoed by others involved in the project.[63]

How would these people learn to program? Well, in a word: constructionism— and they would not need teachers to do it. In a 2006 interview about OLPC, Papert defended the lack of curriculum or teacher support and the expectation that children would teach themselves, stating that millions of people had learned to use and program computers allegedly on their own: "In the end, [children] will teach themselves [how to program]. They'll teach one another. There are many millions, tens of millions of people in the world who bought computers and learned how to use them *without anybody teaching them*. I have confidence in kids' ability to learn."[64] The expectation for children in the Global South to program their own computers (and to fix their hardware) was thus framed not as a burden but as an expression of freedom and individual agency.

This freedom was one that OLPC contributors nostalgically recounted experiencing in their own childhoods. Although Papert admitted that computers had not been part of his childhood—he was born in 1928— he described differential gears as his primary object-to-think-with and explained that these gears had played a role similar to what he wanted computers and Logo to play. Papert described his passion for these gears across several publications, starting with *Mindstorms*. "First, I remember that no one told me to learn about differential gears," he explained; his passion was his alone, and gears were something he explored himself. "Second, I remember that there was *feeling, love,* as well as understanding in my relationship with gears. Third, I remember that my first encounter with them was in my second year [of life]."[65]

Many other OLPC contributors, along with computer programmers across the technology industry, have similarly described learning about computers as something they had done on their own, driven by feelings of passion and freedom and independent of any formal instruction—just as OLPC's intended beneficiaries were meant to do.[66] Shunning school in favor of this individualist learning model, many describe spending hours alone tinkering with Commodore 64s or Apple IIs, or making virtual friends in BBSs or Usenet groups. The ubiquity of this narrative, not just among OLPC developers but across the technology industry, suggests that it holds particular cultural power as a signal that those who share it are technically precocious boys—and that they belong.

However, these narratives of being self-taught invariably smoothed over the social support and scaffolding that, though unacknowledged, had generally helped these hackers. These might have included a stable home environment and a resource-rich, infrastructurally stable community that supported technically precocious children such as themselves, and also often included a parent who was a computer programmer or engineer. OLPC's idea of the self-taught learner who disdains school for computers thus discounted the critical role that various institutions—peers, families, schools, communities, and more—play in shaping a child's educational motivation, technological practices, and identities. The individualism implicit in this story illustrated who they felt the agent of change would be—and took parents, teachers, institutions such as schools, and broader cultural systems more or less out of the equation.[67]

At the same time, OLPC argued that because the software was open source, it could be adapted to match the values of the children involved; they had complete freedom to overwrite any values that OLPC instilled. From the perspective of a group of people who were so familiar with computers, an open-source machine might indeed have seemed like the ultimate value-neutral device. After all, Hill reasoned, if someone didn't like the content, they were free to reprogram it into something they did like. Bender echoed these sentiments on the "OLPC Myths" page of the project's wiki, while pushing back against another writer who wondered if the project might be imperialistic. "OLPC pedagogy is based on Constructionism, the gist of which is that you learn through doing, so if you want more learning, you want more doing," Bender explained. "While this approach is not epistemologically agnostic, it is for the most part culturally agnostic: it is—by design—amenable to adaptation to local cultural values in regard to what 'doing' is appropriate."[68]

Setting aside the ways in which culture and epistemology are not separable, Bender's statement—like the XO's View Source button—presumed a programmer's degree of familiarity with the machine. However, if Hill and other OLPC contributors believed that everybody *should* have that degree of familiarity—they had found great power in such familiarity in their own lives, after all—then the problem of imperialism would disappear: it was the individual's responsibility to learn the skills necessary to reshape the laptop in their own image. Moreover, by making education an individual experience where everybody (in theory) would have all the tools they needed to succeed, it shifted the burden of failure from the system—a flawed educational model, a corrupt government, an unjust economic structure, a missing or crumbling infrastructure—to the individual. If children failed to learn with the XO, when they ostensibly had all the tools they needed for success, it would be nobody's fault but their own.

Trojan Horses, Shortcuts to Change, and the Normativity of Nostalgic Design

At the same time that they claimed the project was value neutral—indeed, sometimes in the same talk or interview—OLPC leaders also said the project was a "Trojan horse." Ignoring the negative implications of the association, both Negroponte and Bender described the laptop as a device that slyly added constructionism to what governments might just view as a basic classroom tool. In a 2005 presentation at the MIT Emerging Technologies Conference, Negroponte said, "With every head of state, I say we are selling you as an idea, a Trojan horse. ... It's a Trojan horse in the form of an e-book."[69] In a 2007 interview, Bender reiterated this comparison with the quote that opened this chapter. Negroponte used the Trojan horse metaphor again in an email he sent to three of the largest public OLPC mailing lists on April 23, 2008, which was reprinted in news articles and elsewhere online: "I believe the best educational tool is constructionism and the best software development method is Open Source. In some cases those are best achieved like the Trojan Horse, versus direct confrontation."[70]

These quotes—along with extensive talk about the "subversive" nature of the machine on mailing lists, wiki discussions, blog posts, and across the web—echo the mistrust of authority and support of innocuous rebellion that are the cornerstones of constructionism and the ethos of MIT's hacker community, as well as of the social imaginary of the technically precocious boy that undergirded both, as we saw in the previous chapter. These sentiments moreover indicate a belief in the power of technology to provide a shortcut to social change: the XO laptop is not so much about training children for particular jobs as it is about liberating and empowering children to think independently, ultimately bringing about peace and an end to poverty. This connects to the description of OLPC's third core principle, in which the computer-enabled education that OLPC would promote is described as inoculating against prejudice: "A healthy education is a vaccination, it reaches everybody and protects from ignorance and intolerance."[71] This belief that the XO and its constructionist software could provide a shortcut to peace and prosperity—even if governments, schools, and parents did not actively recognize its potential for this—reflected a technologically determinist faith in the power of the XO laptop itself to create change. As the "OLPC Myths" page explained, it could provide "the fastest way for poor people to get the political clout to require their governments to provide services to them. Or to get the education for real jobs that take them out of poverty completely. Or access to innovative technologies for providing food, water, clothing, shelter, energy, etc."[72] Moreover, this world changing would be inevitable and natural: as long as OLPC managed to get its laptops out in the world, success would follow.

This tendency toward technological determinism also undergirds Papert's claim that computers as objects-to-think-with have universal appeal. Much of his argument rests on the assumption that given unlimited access to a computer with constructionist software, all children—or at least those he calls the most "intellectually interesting"—will use those programs extensively.[73] Papert claims that computers are more compelling than any previous technology because of their programmability.[74] He asserts that learning to program is much more meaningful to children than existing mathematics curricula because children are "able to bring their knowledge about their bodies and how they move into the work of learning formal geometry."[75] But do children really find computer programming as "fun" as Papert claims they do?[76] Does it have "recognizable personal purpose" for them, as he asserts?[77] Is it really as "concrete" as throwing a ball?[78]

This points to one of the most pressing problems of nostalgic design. On the one hand, it risks excluding those who do not identify with the same cultural imaginaries. Rather than appealing to all children, it might appeal only to those few children who already identify with the same social imaginary of the technically precocious boy as the developers do. One statement on the "OLPC Myths" page points in this direction, stating that the laptop "provides access to education, health, technology, economic opportunity, and more, and a few children will be able pull themselves out of poverty with no other assistance"—never mind the rest.[79]

Moreover, many in OLPC acknowledged—sometimes readily, like Hill— that they were not typical of their peers and, in youth, had often been shunned for their unusual interests or obsessions. They had often been unique in their interest in learning to "think like a machine," even when many others with the same level of access had not been nearly as captivated by computers. More broadly, neither Negroponte nor Papert discussed the possibility that they, and their students at MIT, had enjoyed privileged childhoods, nor did they dwell on what had enabled that privilege. Though Papert used language that encouraged all children to participate in his experiments, his accounts of employing his theories in classrooms and other settings reinforced notions of exceptionalism by focusing his lengthy descriptions on the few engaged children—those rare emblems of "success" who appeared to prove his theories—and not accounting for (indeed, rarely mentioning) the rest.[80] This focus presented children's practices as resulting from innate interests rather than socialization or other environmental factors. It also positioned the child and the laptop as the primary agents, favoring technological determinism—all it takes is the right kind of computer to keep kids as yearners—over the complicated social processes involved in constructing and negotiating childhood and learning.

In their 2003 book, *Unlocking the Clubhouse: Women in Computing*, authors Jane Margolis and Allan Fisher link this essentialist narrative to white, American, male computer scientists in particular. They note that the white, American-born men in the Carnegie Mellon computer science program whom they studied tended both to attribute their success with computers to innate abilities and to tell stories about teaching themselves programming at an early age—embodying the imaginary of the technically precocious boy. Women, immigrants, and those in minority groups, in contrast, were more mindful of what they owed to those around them for their privileged position—and also tended to discover computer science later. Not personally identifying with the imaginary of the technically precocious boy, they instead developed oppositional subject positions and drew on alternate motivations for going into the field.[81]

This disconnect between the narrative of personal idiosyncrasy that is common in computing cultures and the mission of OLPC to change the world may seem incommensurate: how can these projects expect to change the world if the kind of child they are trying to reach has been formed in their own rare, culturally bounded image? Why would those involved with the OLPC project believe that a social imaginary that only a very narrow segment of the American population personally identifies with would generalize to entire populations of children across the Global South?

Indeed, rather than appealing to the "natural" state of *all* children, OLPC was actually quite specific about the kind of child—and the kind of technology—that fit their vision for changing the world. In his descriptions of "yearners" and "schoolers," Papert entirely left out many potential users—from teachers to children with diverse interests or those lacking the requisite technical expertise—and a range of potential uses. Instead, he— and OLPC—designed the XO to support technical interests and ascribed those interests as innate. Similarly, OLPC's developers, some of whom told stories of being marginalized in their own youths but of finding power through hacking and programming, wanted to translate this redemption narrative to the Global South without fully recognizing the social, institutional, and infrastructural factors that were also part of the story—and the way that their own subject position informed their telling of it. A likely result—as we will explore in the latter chapters of this book—was that this translation would fail, that children in the Global South just would not see a vision for themselves and their futures through laptop use.

On the other hand, perhaps the project's objective was more normative: to *create* the kind of child it wanted to see in the world. This is not only about the design of idealized technologies but idealized social worlds: worlds that assume a particular set of competencies and possible social changes, assumptions that are then inscribed into the technology itself and its expected user.[82] Nelly Oudshoorn, Els Rommes, and Marcelle Stienstra found that although attempting to design for "everybody," the predominantly male designers they studied had inscribed their own tastes, competencies, and views of gender identity onto their users in an attempt to "configure" these users in the designers' own image (what the authors called the "I-methodology").[83] The result was a "masculine design style" characterized by the designers' interests—much as nostalgic design had resulted in a design style that referenced designers' idealized pasts and the social imaginaries that had shaped these memories and identities. The I-methodology thus resembles nostalgic design, though nostalgic design more specifically tries to evoke past technologies, social worlds, and collectively held social imaginaries—imaginaries that obscure this normative mission with a sense that children are naturally this way.

The practice of nostalgic design—of designing for imaginaries of childhood, here the technically precocious boy that rebels against school and finds solace in computers—is thus a political act. Papert, Negroponte, and other OLPC leaders did not critically consider whether their experiences really could be universalized. Moreover, they ignored the very different conditions in which different childhoods play out and took technical creativity as innate rather than learned. For these reasons, it is critical to recognize the disconnects between social imaginaries of childhood and the real-world practices they (try to) configure. OLPC contributors drew on a shared set of cultural ideals about childhood when they discussed the kinds of individualism, creativity, and empowerment that they thought their laptop could unlock in children around the world—and their design configured the kind of child they expected to encounter. What seem like purely technical resources—such as access to a specifically designed computer— are in fact social and political in nature, embedded in histories of learning and childhood that constitute much of how we envision technological development today.

Collectively, the promises of play, connectivity, and freedom that the XO laptop embodied—along with the faith that the laptop would be a shortcut to the changes this group wanted to see in the world—served as rallying cries within the One Laptop per Child community and a beacon for outsiders who wanted to contribute time or resources to the project. Without these themes and the project's mission, the XO was a cheap, underpowered laptop. But with the features that embodied these themes—the playful appearance, the games and game engines, the mesh network and security suite, the open-source software and View Source button—it became suffused with heady potential. Its lackluster performance stopped being a liability; it was either absent from discussion or was discussed as a nostalgic throwback or power-saving feature.

Still, the reliance on social imaginaries of childhood through the practices of nostalgic design underscores an ideological slippage between the ideas taken up by XO laptop designers and the lived experiences of the children they hoped to reach—experiences we will spend the rest of the book exploring. Locating these slippages, and the tenacity with which supporters held to their ideals even when confronted with them, provides a powerful method of identifying charisma and uncovering its possible consequences.[84] The next chapter will begin to examine just what became of this Trojan horse when it was put into use. The original Trojan horse story, after all, did not end well for Troy.

3 Translating Charisma in Paraguay

We'll take tablets and drop them out of helicopters into villages that have no electricity and school, then go back a year later and see if the kids can read.
—Nicholas Negroponte, quoted in Thompson, "Negroponte Plans Tablet Airdrops."

For many in Paraguay, 2008 was a year of optimism. In April, liberal candidate Fernando Lugo won the country's presidential election on a message of hope, much as Barack Obama would do in the United States that November. With 43% of the vote, compared to the 31% of his opponent (who was the first woman presidential candidate in Paraguay's history), his win marked the first time that Paraguay's conservative Colorado Party had relinquished power in sixty-two years and was one of the few peaceful transfers of power between parties in the country's two-hundred-year history since independence.[1] Known by his supporters as *Obispo de los Pobres* (bishop of the poor) for his humanitarian service as a Catholic priest in one of Paraguay's most destitute districts, Lugo centered his campaign on reducing Paraguay's bleak social inequalities and widespread corruption.[2] Lugo's messages of hope and reform lasted beyond his election: during my fieldwork two years later, participants told me about their newfound hopes for Paraguay's future on the world stage, even as Lugo became embroiled in scandal and faltered in implementing the dramatic changes he had promised.[3]

Alongside Lugo's ascent, a young, well-connected, and idealistic Paraguayan took up this banner of hope and steered it toward technology-enabled learning. Raúl Gutiérrez, then a student at the Catholic University in Asunción, first started exploring the idea of OLPC in Paraguay in 2006.[4]

He pitched the idea in April 2007 to Luis Castiglioni, who was the father of a university friend and at that point a candidate for the Paraguayan presidency. Castiglioni was so excited about OLPC that he named Raúl, then 24, his honorary technical advisor and sent him to an OLPC country workshop in Boston in April 2007, where Raúl met the OLPC team. However, Castiglioni's candidacy did not last through the presidential primary (he later became Lugo's vice president), and the possibility of a government-backed OLPC project faded with Castiglioni's prospects.[5] Still, Raúl maintained hope. In 2008, around the time Lugo took office, he joined forces with Cecilia Rodríguez Alcalá, who had just graduated from Tufts University in the suburbs of Boston, not far from MIT. Cecilia had visited the MIT Media Lab during her time in Massachusetts, where she learned about the One Laptop per Child project and was similarly captivated by its promises. Buoyed by news of large-scale OLPC projects underway in Peru and Uruguay, the two founded a nonprofit, nongovernmental organization (NGO) called Paraguay Educa to bring OLPC's distinctive laptops to their country.

They had some powerful allies in their quest. *ABC Color*, one of Paraguay's two major newspapers, contributed good press as well as office furniture to the fledgling NGO, which settled into an estate turned office in the suburbs of Asunción provided by a relative. A September 2008 article in *ABC Color* explained Paraguay Educa's motivations as closely aligned with OLPC's, praised constructionist learning, and extolled the benefits of OLPC's five core principles of child ownership, low ages, saturation, connection, and free/open-source software for Paraguayan children. The article described how the laptop's customized learning software, rugged construction, low power consumption, and state-of-the-art screen would let children leapfrog past Paraguay's anemic educational system and become adept programmers, mastering the mathematical thinking that is especially valued in computer engineering.[6]

This story of charismatic technology is a familiar one. It adopts the same worldview and buys into the same social imaginaries as OLPC itself did— and makes the cultural change its laptop promised not only compelling but seemingly effortless, natural, even inevitable. In fact, OLPC founder Nicholas Negroponte often characterized duties of OLPC and local partners as ending when laptops were handed out. At the Techonomy conference on August 9, 2010, he said, "One the things people told me about technology, particularly about laptops in the beginning, 'Nicholas, you can't give a kid a laptop ... and walk away.' Well you know what, you can. You actually can." He went on to characterize the kind of change he thought would be possible from a project run this way, although he did not offer evidence to back his claims: "And we have found that kids in the remotest parts of the world ... not only teach themselves how to read and write, but most importantly, and this we found in Peru first, they teach their parents how to read and write."[7] Although not everybody in OLPC subscribed to this degree of technological determinism—a 2013 book on OLPC that founding member Walter Bender cowrote, for instance, discusses the importance of local partnerships, which we will explore more in this chapter and in chapter 6—it still characterized the process of cultural change as simply a matter of lining up the right resources in order to let the laptop speak for itself.[8]

Even though this charismatic story did important work in mobilizing support in Paraguay and around the world, the on-the-ground work that was needed to implement various OLPC projects challenges this vision. This chapter focuses on the work that was required to *translate* this vision through a medium-sized project of ten thousand XO laptops in Paraguay, which is considered by many in the OLPC community to be one of the most successful.[9] I use "translate" here not only to suggest a process that requires knowledge of two worlds and work to bridge them—what is needed, for instance, to translate from one language to another—but also to provide a framework for discussing the contexts of the various actors involved in this process. In Paraguay, these actors include OLPC, Raúl and Cecilia's NGO Paraguay Educa, local and national governments, school directors, teachers, parents, the laptop, and of course children—though children will be the focus of the next two chapters. Leaning on how "translation" is understood in science and technology studies, this lets us better understand where and why the work of translation succeeded—and where and why it failed.[10] In fact, we will see that due to infrastructural and labor challenges, OLPC's XO laptop was not able to "speak for itself"; it required extensive work to achieve even partial translation. We will also see what Paraguay Educa and Paraguayan teachers did to (try to) overcome these challenges. In the process, this chapter shows that the charismatic promises of computer-aided learning and child-driven cultural change were brittle—easy to break and labor-intensive to repair and maintain—when put into practice.

OLPC in Paraguay and across Latin America

OLPC's early publicity stated that the organization would only sell XO laptops to governments in the Global South and in lots large enough to provide one for every primary-school student in the country. However, the price of the XO never dropped below $188 per unit, and few countries could afford even this level of expenditure on education, much less what proved to be substantial setup and maintenance costs, as we will see.[11] OLPC thus gradually decreased their large purchase requirements. Moreover, even though much of OLPC's early publicity referenced African children as its imagined beneficiaries—with promotional pictures featuring groups of smiling African children on the laptop's screen—less than 10% of XO laptops in use around the world are in Africa, almost all of those (some quarter of a million) in Rwanda.[12] Instead, projects in Latin America account for nearly 85% of the XO laptops purchased around the world. By 2018, Peru and Uruguay had purchased around one million laptops each, together accounting for more than three quarters of XO laptops in existence. Paraguay's project is not on this scale, though its scale is still much larger than a pilot of a few dozen or few hundred machines. It joins other medium-sized, regional projects in Argentina (La Rioja), Mexico, Mongolia, the United States (Birmingham, Alabama), Haiti, and Nicaragua with between ten thousand and sixty thousand XO laptops.[13]

The support for OLPC in Latin America fit with regional goals and identities. A number of leaders and government officials across Latin America attended MIT or were familiar with the Media Lab and its projects, as Anita Say Chan found in her work on OLPC and the techno-cultures of Peru.[14] OLPC leaders including Seymour Papert, Negroponte, and Bender referenced Brazilian educational philosopher Paulo Freire—who, in *Pedagogy of the Oppressed*, called for a radically new model of education that did not subjugate the lives and worldviews of students—in order to justify calls for child-led "learning by doing" instead of schools, subjects, and set curricula, though these references did not discuss the more radical elements of Freire's proposals, such as his connections between literacy, critical praxis, and emancipation.[15]

Moreover, OLPC's commitment to open-source software connected well with an intragovernmental push to promote open-source software across Latin America. Even though many open-source projects originate in the

United States—and those that do not are often still oriented toward US-based computing cultures, as Yuri Takhteyev found in his ethnography of the Brazil-based developers of the Lua programming language—the narrative of open-source software being value neutral was one that resonated not just with OLPC but with many across this community in Latin America, and it also fit with visions of regional self-determination over foreign dependence.[16]

These reasons resonated among the founding members of Paraguay Educa as well—as did the imaginaries on which OLPC's charisma was based. In particular, cofounder Raúl and his first employee, Martín Abente, were both skilled programmers, among only a handful of programmers in local projects to contribute code "upstream" to the main Sugar software build. Raúl had also contributed to other open-source projects while in college, before he heard about OLPC. They both identified as hackers and, like many of OLPC's employees and contributors, wanted to bring the freedom and joy they had experienced in this identity to other children. They were, of course, Paraguayan, but their experience of being Paraguayan differed markedly from the experience of most of the country's population. They were part of the small cosmopolitan elite in Asunción accustomed to global travel, and they spoke English much better than Guaraní, the second official language of Paraguay and the only language spoken in many rural areas. In addition to a passion for OLPC and open-source software, they loved watching anime, made nerdy technical jokes, and referred to one another by their online handles—all of which gave them common ground with other programmers around the world. Assisted by visits from two former OLPC staff members, their technical team became one of the strongest of any OLPC project: they wrote software updates for several components of Sugar, including the web browser and the security system; wrote software for the school servers; and developed a web-based inventory software package to track which laptops were given to which students, which several other OLPC projects subsequently used.[17] OLPC projects in other countries, as well as OLPC itself, later contracted with Martín to continue this work.[18]

Paraguay Educa sought funding for its first batch of XO laptops throughout the second half of 2008—a task that proved difficult. Despite Raúl's early contact with OLPC and support from Paraguay's vice president, the NGO was turned away by Paraguay's new government, as well as by OLPC itself because it was still pursuing country-wide projects at the time.

But their luck turned when they secured funding from the philanthropic arm of Swift Group, a European banking conglomerate, to purchase four thousand first-generation XO laptops from OLPC, which arrived in early 2009.[19] The NGO distributed these laptops in April 2009 to all students in first through sixth grades (ages six through twelve) and their teachers in ten schools in the small municipality of Caacupé, fifty kilometers east of Asunción.[20] In 2010, with funding from the Inter-American Development Bank and the philanthropic branch of Itaipú Dam (until 2012 the world's largest hydroelectric dam, of which Paraguay owns half), Paraguay Educa purchased another six thousand laptops and, in May 2011, gave them to all primary-school students and teachers in the other twenty-six schools in the municipality, as well as new first- and second-grade students in the initial ten schools. With every primary-school student in the area owning a laptop, Paraguay Educa hoped that Caacupé would serve as a proof-of-concept city and that Paraguay's government would then take the project nationwide.

Though run by an NGO, a project such as Paraguay Educa's, which proposed to reform education so deeply, could do so only with the support of the local and national governments responsible for overseeing public schools—support that was enabled by Raúl's and Cecilia's connections. When their needs were well aligned, this generally worked out. For instance, Paraguay Educa wanted to link XO laptops to individual students. This required that all students be uniquely identifiable, which was possible via their *cédulas* (national ID numbers)—except that many children did not have one. Officially, parents were supposed to register for their child's *cédula* when the child was born, but many did not, especially in rural areas.[21] Caacupé's municipal government was excited to use the laptop program as a chance to force those parents who had not registered their children with the state to do so.[22] Paraguay Educa and the local government framed the requirement to have a *cédula* as part of "being a good citizen" (it was needed to vote, for instance).[23] Yet it could also allow the state to more easily track a citizen's taxes and fines, enforce the required year of military service for men, and potentially suppress dissent. These concerns were especially salient because just a few decades earlier, Alfredo Stroessner's Paraguay, like Augusto Pinochet's Chile and Jorge Rafael Videla's Argentina, had been a repressive dictatorship that abducted and murdered citizens deemed troublesome.[24] In the 2011 Phase II laptop distribution, this requirement was extended from just *cédulas* to both *cédulas* and current vaccinations.[25]

Aside from its cooperation around *cédulas*—and in spite of its contacts in the national government—Paraguay Educa's relationship with the local and national government was complicated and sometimes conflicting. Even as other municipalities throughout Paraguay expressed interest in being part of future phases of the project (interest that did not pan out in the end), the municipal government of Caacupé supported the project verbally but often resisted backing its words with resources.[26] Similarly, the national government praised the project but did not always back that praise with action. In September 2008, President Lugo agreed to support Paraguay Educa's cause in exchange for a commitment to teacher training, exchanges with other Latin American OLPC projects (particularly Uruguay's), and a road map for rolling out the project nationally.[27] Lugo later stated that he hoped to provide laptops to at least half of Paraguay's children during his term in office.[28] However, not only were these promises empty, but Paraguay's Department of Education and Culture at times proposed competing with Paraguay Educa by deploying its own technology initiatives, tied to its members' own legacies or reelection campaigns.[29] Thus, the Paraguayan and Caacupé governments ultimately did not provide many resources to help the NGO in its mission, much less take it over and grow it.

Uruguay's OLPC project provides a sharp contrast to Paraguay's in terms of resources, though they bear a number of similarities otherwise. Unlike Paraguay's NGO-run project, Uruguay's government-run project was guaranteed long-term funding through a change in Uruguay's tax code, which set aside a sizable portion for ongoing support of educational technology initiatives. Uruguay, a small coastal country nestled between Argentina and Brazil with a population of just 3.4 million in 2010—about half of Paraguay's population of 6.3 million in the same year—is classified as high income by the World Bank, has a strong educational system, and consistently boasts over 98% literacy, the highest in the region.[30] Plan Ceibal, the governmental division in charge of Uruguay's OLPC program, was created in 2008 by president Tabaré Vázquez to distribute 450,000 laptops to primary-school students (aged six to twelve) and their teachers by the end of 2009, starting with the outlying areas of Uruguay and ending in the capital, Montevideo, where about one-third of the country's population lives.[31] In 2010, the project was renewed by new president José Mujica, who added another 550,000 laptops throughout his five-year term for incoming primary-school students as well as for high-school students who were too old to receive one initially. In 2015, Vázquez was again elected president and continued to support the project, making Plan Ceibal one of the few OLPC projects still active and fully supported in 2018.

In contrast, landlocked Paraguay, nicknamed *el corazón de America del sur* (the heart of South America) for its continental centrality, is a conservative, Catholic country with a weak civil society, spotty infrastructure, and a poor record for women's rights. It is much poorer than its neighbors Uruguay and Argentina, and its wealth is unevenly distributed.[32] The World Bank estimated that in 2009, the year Paraguay Educa distributed its first batch of laptops, 35% of the Paraguayan population lived in poverty, with 19% (mostly rural subsistence farmers) in extreme poverty, though this represented an improvement since the overthrow of Stroessner in 1989, after thirty-five years of dictatorial rule.[33] During Stroessner's dictatorship, education was systematically defunded and devalued: when he was overthrown, only 1.7% of the country's gross domestic product was spent on education, including university education, and this number had only risen to 3.8% by 2010, the year I arrived in Paraguay to observe the project (still lagging behind the Organization of Economic Cooperation and Development average of 5.4%).[34] The official minimum wage in Paraguay in the second half of 2010, during my fieldwork, was 1,507,484 guaranies (about $320) a month.[35] Minimum-wage laws were underenforced, however, and the majority of the population worked either in the exempt public sector (including teachers) or in Paraguay's extensive informal economy, many of them making significantly less than minimum wage.[36] Despite creating challenges, these conditions also made Paraguay a more ideal intervention site as imagined by OLPC leadership: a place of poverty, in need of educational reform.

Paraguay Educa's Infrastructural Translations

Despite the fact that Paraguay Educa believed in the charismatic ideals of One Laptop per Child, the NGO departed from Negroponte's claim that it could just hand out laptops and walk away. Children would need to be able to charge their computers and connect to the internet, and neither capability was guaranteed in schools or homes. So, like the much-larger OLPC project in Uruguay to the southeast, Paraguay Educa supplemented its two laptop handouts with substantial infrastructural support at every school, the child-frequented space that it could most easily arrange to access.[37] With the help of corporate donors and community volunteers, it installed WiMAX antennae on top of tall rebar towers, school servers and wireless access points in metal cages, and power outlets in classrooms before laptops were given to the school's students. It also installed Wi-Fi in the main public plaza of Caacupé in Phase II of the project.[38] The need for these infrastructural interventions, which were fairly expensive and labor-intensive, already begins to destabilize OLPC's narrative of easy cultural change.

Notes from my visit to one of the ten Phase I schools in early September 2010 illustrate many features of a typical Paraguayan primary school, as well as these infrastructural changes. When I arrived just before school started that balmy spring morning, children were congregating in the school's concrete courtyard, surrounded on three sides by classrooms and trees flowering in a profusion of colors after a particularly cold winter. These classrooms, as in other Caacupé schools, had a concrete floor, wooden rafters above, and stuccoed brick walls painted light green. These walls were covered with decorations: large paintings of clowns and cartoon animals, pictures of Jesus, student projects, and various posters depicting things such as the muscles of the body, Guaraní letters, Spanish parts of speech, how to care for the environment, and—a new addition—how to care for an XO laptop.[39] Classrooms had lights and, at least in this school, also had the powerful ceiling fans common in wealthier Paraguayan houses, which helped mitigate the steamy temperatures in late spring and early fall. Paraguay Educa, the municipal government, and parent volunteers had collaborated to install one power outlet in each classroom for recharging laptops.

One third-grade classroom had its eighteen well-worn wooden desks organized in rows facing the blackboard. In the classroom next door, desks were instead arranged in a U shape, while in a sixth-grade classroom across the courtyard, students sat around a large wooden table. A teacher's desk, generally stacked high with papers and notebooks, sat in a corner of each classroom. The wooden doors and window shutters were kept open during the school day. Some of the glass panes in the metal-rimmed, louvered windows were broken, but the weather was generally so warm, except for perhaps a few weeks in July (coinciding with winter break), that it did not really matter.

Some teachers were still writing up the day's lessons on their classrooms' chalkboard: one section of the board for language, one section for mathematics, one section for natural science, and one section for a weekly or semiweekly lesson on a fourth subject, such as social science, health, arts, or culture, as required by the countrywide curriculum set by the Ministerio de Educación y Cultura (Ministry of Education and Culture, or MEC). This school, like most of the larger public schools in Caacupé, did have a few textbooks for language arts and mathematics from the MEC but not for all grades, and the paperback books were over a decade old and well worn. As a result, these books were only used for reference, and teachers wrote lessons on the chalkboard for students to copy and complete in their notebooks, with varying degrees of interactivity. Students supplied their own notebooks and pencils, both of which could be purchased at the school cantina.

On one edge of the courtyard was the school's central office—a simple room, with desks for the school director and coordinator, a few chairs, and some shelves for school records. The office doubled as a storage space, and the edges of the room were stacked high with old, worn textbooks and cardboard boxes of ultrapasteurized milk in Tetra Paks from the municipal government to give to the students after recess. The office also hosted the school server that Paraguay Educa had installed, locked in its metal cage. Above it was a printed list of the prices for replacement parts for the XO laptop. Right outside was another piece of infrastructure from Paraguay Educa: a narrow tower of welded steel trusses, painted red and white and not much wider than a flagpole, that elevated the school's WiMAX antenna to connect it to the internet via bandwidth donated by Latin American telecommunications company Personal Telecom. Bolted to the ceilings of the school's open-air hallways were wireless access points provided by Paraguay Educa, in metal cages like the school server, which spread this signal throughout the school.[40] Conspicuously absent—at least to my privileged eyes—were many other technologies common in primary- and secondary-educational settings in the United States: there were no projectors, spare or loaner laptops, other computers, televisions, printers, photocopiers, paper for school use, or a bell or intercom system. Neither the student nor the faculty bathrooms had soap or toilet paper, and they were frequently out of service.

This description of this public school in Caacupé—which was not the wealthiest in the area but was also far from the poorest—illustrates the extent to which Paraguay's infrastructure was lacking. It also challenges some of the assumptions that OLPC's leadership made about schools in the Global South. The first is the absence of textbooks. During his 2006 talks, Negroponte often billed the laptop as equivalent to a textbook, which could be provided on the device. Amortized over five years, he claimed, a hundred-dollar laptop would be equivalent to the twenty dollars per year per student that Brazil and China budgeted for textbooks.[41] But Paraguay certainly did not spend twenty dollars per year per student on textbooks; few public schools used textbooks in classrooms at all.

Taking the analogy even further, Papert equated laptops not to textbooks but pencils: to him, a laptop was an object-to-think-with as basic as a pencil, so having one computer per classroom would be as ridiculous has having one pencil per classroom. This parable also appears in OLPC's mission statement. "One does not think of community pencils—kids have their own," it asserts. "A computer can be the same, but far more powerful."[42] However, computers are still several orders of magnitude more expensive than a ten-cent pencil—and even a ten-cent pencil was not something all Paraguayan students could consistently afford.

We have thus already seen some cracks in OLPC's focus on individuals and computers, as well as in the assumptions it made about what a project would need to provide: Paraguay's physical infrastructure needed fairly substantial upgrades to enable connected laptop use, which added considerable costs on top of the baseline expense of the XO laptops. Still, much of this infrastructure was a one-time cost. What happened after this was in place and laptops had been given to children and teachers?

Trojan Horses, Breakdown, and Disruption

One Laptop per Child had hoped to disrupt traditional classroom culture—Negroponte, Bender, and others in OLPC's leadership referred to the project's XO laptop as a "Trojan horse."[43] How did this disruption play out in Paraguay? To find out, we can turn to a description of another school from my field notes. It was midmorning, just after recess, on a cool winter day in August 2010, and I was observing a fourth-grade public school classroom in Caacupé. The sun peered weakly through the thick smoke from nearby farmers burning their fields in preparation for the planting of next year's crop, casting two orange parallelograms on the cracked tile floor.

The fourteen students in attendance that morning haphazardly faced the board from the classroom's sixteen small wooden desks, well worn and carved with layers of graffiti. These were only about half of the students who would use this classroom today. This school, along with all other schools that I observed except one, hosted two student bodies. Children in the morning session attended school from seven to eleven in the morning (or sometimes eight to noon), with a half-hour break for recess halfway through the morning. Then these children went home for lunch and the rest of the day, and another set of children came after lunch for the second school session from one to five in the afternoon, with a half-hour recess at three o'clock. Schools did not serve any meals, though larger ones had a small cantina where students could purchase candy and snacks.

This morning, these students were already halfway through their four-hour school day. Before recess, they had finished their second lesson, on mathematics, using the most common tools for education in Paraguay: pencils and slim paper notebooks. The third lesson was on health, and they would be using XO laptops. After handing back the homework in students' notebooks that she had graded during recess, the teacher told her fourteen students to start up Tux Paint, a popular, open-source drawing activity. Nine children took out their laptops and started them up, but five did not. One girl sat in the corner, staring into space. When I asked where her XO laptop was, she said it had a broken screen. Another girl without a laptop had her notebook out. The teacher pointed out one boy who had transferred here from a school not in the program and thus had never received an XO. Two other boys without laptops told me, "Se rompió" (It broke), implicitly dodging responsibility with the blameless Spanish reflexive. These three boys huddled around the laptop of a fourth boy, watching everything he did.

"I don't want to hear that nobody has it," the teacher said of Tux Paint, the software she hoped to use. But when the teacher and I checked the laptops one by one, only two of the nine students with XO laptops had Tux Paint installed. "We used this activity last week!" the teacher admonished. But she had prepared for this moment: she got out her USB flash memory stick (often called a "pen drive" in Paraguay), announced that it had a copy of Tux Paint on it, and passed it to the nearest student. He plugged it in, navigated to the USB directory, and clicked the Tux Paint icon—and his computer froze on the first installation screen. I jumped in to help. After force-rebooting the first laptop, we cleared space on several laptops that did not have room on the hard drive for the installation.

Because of the stamps and sound files, Tux Paint was a large activity to install, and this process did not go quickly, even with only seven laptops to manage. At this point, most of the children in the class had stopped paying attention, and several boys were horsing around on the left side of the room. Their roughhousing jostled one of their laptops, which was covered with stickers of cars and anime characters, from the small attached desk. It hit the concrete floor, handle first, with a loud *thwap*, and the whole room gasped and turned to stare. The owner of the fallen laptop gingerly picked it up, examined it, and declared, "¡Anda!" (It runs!), to a collective sigh.

While we were attending to installation, two of the laptops that had been turned on since at least the beginning of class ran out of battery power. Their owners shut them and plugged them into two chargers dangling from the class's two power outlets, near the door of the room. But then another XO ran out of battery power, and there was no outlet available to charge it. The student reluctantly closed it, took out his notebook, and started copying from the board. Other laptops' battery indicators showed orange—they were starting to get low—but were not yet red. Forty-five minutes after the teacher had initially told everyone to take out their XO laptops and start Tux Paint—nearly one-quarter of these children's instructional time that day—the six laptops that were still running finally had Tux Paint installed. The final ratio of running laptops to students was less than two to one.

The narrative above shows that the XO laptop was a different kind of Trojan horse in the classroom, one that hijacked lessons with chaos and technical difficulties. This was due to the design limitations of the machine itself, the charismatic but ultimately unrealistic promises that OLPC had made about its robustness and battery life, and OLPC's early disregard for the on-the-ground realities of teachers around the world. These issues produced a lot of friction in use, particularly in the classroom. The focus on individual learning in the place of, or even at the cost of, institutional support indeed disrupted these institutions and those who were part of them, but the kind of disruption that occurred was much less about shifting paradigms than introducing breakdowns.

1. Power and Batteries

The vignette illustrates three particularly thorny issues: power, software, and breakage. Negroponte boasted in 2005 that the then unfinished XO laptops could be charged by an integrated hand crank and even showed off a non-working mock-up with this crank at his presentation in Tunis in November 2005. In practice, however, the hand crank was never included on later working prototypes or on any production machines because it would have put too much torque on the laptop's innards and likely too much strain on the small muscles in its users' forearms.[44] Though the power-management hardware and software were designed to accept flexible sources of power, all XO laptops I have encountered or heard about around the world have been charged by plugging the conventional AC adapters that shipped with the laptops into wall outlets.[45]

In theory, charging should have been particularly easy in Paraguay, especially compared to other OLPC sites such as the remote highlands of Peru. Cheap and plentiful, electricity in Paraguay is provided by the gigantic Itaipú Dam—which has the second-largest capacity in the world—and two other dams, and it reaches between 89% and 95% of households.[46] Even so, the electrical infrastructure in Paraguay was unreliable and blackouts were common during my fieldwork, especially during heavy rains. Moreover, not all rooms had power outlets; in fact, parents, teachers, and Paraguay Educa had to retrofit school classrooms with outlets at the beginning of the project, and many of the houses I visited had only a few outlets total, most of them in use by lamps, radios, televisions, and appliances.

Early in the project, OLPC leadership had made grand promises that XO batteries would last many hours, even days.[47] As a result, Paraguay Educa did install some outlets in classrooms but did not initially foresee the need for more than one or two outlets per room. Even though the XO used less power than a typical laptop, its battery charge did not last much longer. For typical (internet- and media-heavy) use, even laptops with new batteries would last a few hours at most. In the year and a half between when students received their laptops in April 2009 and my fieldwork in 2010, some batteries had degenerated enough that they ran out of power in less than an hour. The one or two outlets that Paraguay Educa had thought would be sufficient soon became a major limitation. Even if a teacher had access to power strips to daisy-chain across the classroom—which I only saw at one of the wealthier private schools in Caacupé—there were still rarely enough AC adapters that weren't broken to charge all of the laptops at once.

Paraguay Educa and Caacupé schools tried to adapt to address these challenges. Paraguay Educa's technical team built several bulk charging blocks, which could each charge a dozen laptops simultaneously—but it did not have enough of these for every school to have one, much less every classroom, and did not have the resources to mass-produce them. Between my 2010 fieldwork and 2013 follow-up, one school installed dozens of outlets, at desk height, all around several of its classrooms. However, most of these outlets had failed within a few months because the wire used was incorrectly labeled and was not actually rated for the task—an expensive mistake for a poor school.

2. Missing Software

Another frustrating and time-consuming barrier to classroom use, as we saw above, was missing software. To aid in the self-guided exploration that it hoped children would do, OLPC's fifth core principle, "Free and Open Source," advocated for using only open-source software on the laptop. This way, OLPC's leaders claimed, technically precocious children could not only use the software but eventually learn to change it themselves, embodying constructionism's focus on "objects-to-think-with" that have a "low floor" (they are easy to start using) and "no ceiling" (there is virtually no limit to the learning experience).[48] Together with child ownership, this policy attempted to decouple the learning process from the classroom and the teacher by providing children with increased agency over their experiences on the laptop.

Although unrestricted access to laptop software in itself could be benign for most students and possibly beneficial for a few tinkerers, it became particularly problematic in combination with four physical limitations of the XO laptop: its small hard drive (a one-gigabyte solid-state drive in the first-generation laptop), its unexceptional battery life, bugs and instability in the Sugar software, and the slowness of the shared internet connection. The smallness of the XO laptop's hard drive in particular meant that children who downloaded music, movies, games, and other content quickly ran out of space on their hard drives, pushing them to figure out how to uninstall software. In a way, children were doing what OLPC had hoped for—they were customizing their laptops—but most were later at a loss for finding and reinstalling missing activities when it came time to use them in class. This put the burden on teachers to manage the process, complicating their duties by introducing unpredictability and usurping their agency in the classroom. Then, reinstalling missing software was not helped by the slow internet connection nor by laptops crashing during installation. Whereas software developers know that in practice no software is bug free, I found that many teachers lacked this insight into software development and became frustrated when an XO behaved unpredictably—especially when the time they had for laptop use was already quite constrained.[49]

The most straightforward solution to uninstalled software would involve disallowing uninstallation of certain activities, but this would also compromise OLPC's core principles by restricting children's free access to the laptop. While I was in Paraguay in 2010, I was privy to an ongoing debate within Paraguay Educa about designating "core" activities that would be much more difficult to uninstall in future releases of Paraguay Educa's branch of Sugar: the education team unanimously advocated for the change, and the software-development team, who would actually implement the change, equivocated, citing a desire to keep OLPC's core principles intact. When I visited again in 2013, it seemed that the technical team's ideals had prevailed; the XO laptop's software was still completely open.

3. Breakage and Repair

The narrative also illustrates that these laptops, though ruggedized, were not immune from breakage: three out of fourteen laptops belonging to students in the room were unusably broken. This proportion roughly matched what my August 2010 snapshot of Paraguay Educa's inventory and repair database showed: just over one-quarter of Paraguay Educa's laptops had documented but unfixed hardware problems. Of the 1,095 unfixed hardware problems in Paraguay Educa's Inventario (inventory) system that August—roughly fifteen months into the deployment—474 involved a broken charger, 403 a broken screen, 139 a broken keyboard or trackpad, and 79 other hardware issues.

This breakage proportion dwarfed the 1% of extra laptops that, according to Walter Bender, shipped with OLPC orders to account for what were supposed to be rare hardware problems.[50] This problem appeared to be compounded by OLPC's claims and demonstrations of the laptop's ruggedness. In early presentations, Negroponte and others in the OLPC leadership would fling a closed XO laptop across the stage, then pick it up and start it successfully.[51] However, students with broken laptops reported to me that their laptops were often broken not when they were closed but open. Because of OLPC's early claims regarding the XO laptop's ruggedness, Paraguay Educa and other deployments, including the countrywide deployment in Uruguay, did not initially stress the importance of being careful with the laptop.

In my field observations, I noted that the ubiquity of concrete or tile floors and rough cobblestone roads was especially problematic in Paraguay, and many of the breakages occurred from falls. On multiple occasions, I witnessed children running across the concrete schoolyard or down a cobblestone road with an open XO on their arm. One child with a broken laptop told me of using his XO on his bunk bed and accidentally dropping it a meter and a half to the cement floor below. Others put their laptops on a high shelf or on top of a wardrobe, only to have it jostled off. One child told me in an interview that she lent her laptop to her younger brother, who was too young to have an XO himself, and it was returned with a screen that no longer turned on.

Several flaws in the XO laptop's final design exacerbated these problems.[52] Though the XO had a solid-state hard drive that could not head-crash in the way that a rotating platter hard drive could and was relatively well sealed against dust and liquid, especially when closed, the laptop's screen in particular was not resistant to direct impact.[53] At the same time, the screen-side location of the XO laptop's built-in camera invited children to walk around with the laptop open, and the antennae were easy to grab but not very secure as handles. As a result, broken or black screens were the second most common hardware issue, after broken chargers, and more children with still-working screens had fingerprint-sized areas of forever-black pixels, dead from impacts.[54]

There were other problems associated with the first-generation XO machine. Its membrane keyboard was slightly too thin to withstand heavy use, and over time, it cracked at the edges of the most used keys, inviting fidgety fingers to pick at them further. Similarly, the flanges between the charger cable, AC adapter, and computer plug could not withstand swinging or repeated pulls out of the wall from the cable—or even more gentle use over many months, as I also discovered with my own AC adapter. The trackpad that controlled the mouse, located below the keyboard, was notoriously unreliable and finicky.[55] Even though children learned how to reset it by holding down the keys at the four corners of the keyboard and counting to ten, many still complained of a jumpy or frozen *flechita* (cursor; literally, "little arrow"). Thus, though OLPC's designers anticipated some potential hardware problems, their lack of longitudinal field-testing with their target population missed other problems before the launch of the first-generation XO laptop.[56]

The obvious solution to the problem of breakage involves more resources: extra laptops for school use, a consistent source of repair parts, and perhaps money to subsidize the cost of repair parts in cases of financial need, much as Uruguay's Plan Ceibal project had.[57] When Plan Ceibal started noting similar levels of breakage in Uruguay, they contracted directly with the laptop manufacturer, Quanta, for spare parts. A nationwide technical-support network repaired broken laptops, installed new software, and installed and maintained the electrical and telecommunications infrastructure to support the project—a network further supplemented by a group of volunteers called Ceibal Red (Ceibal Network)—though even with this extensive support many laptops reportedly went unfixed.

Responding to this challenge, Paraguay Educa tasked the team that it had hired to install school infrastructure with also repairing and maintaining laptops. During my 2010 fieldwork, the team balanced infrastructural installations at Phase II schools with a rotating schedule of visits to Phase I schools, generally visiting each school once a week. During these visits, it documented problems and fixed what it could.

Software problems were generally easy to resolve, but hardware problems were not: Paraguay Educa had difficulty procuring repair parts, and these parts were expensive. According to Paraguay Educa's technical staff and also mentioned in the book by Bender and colleagues, OLPC maintained that not selling repair components was part of its policy of not handling the maintenance of deployments itself and under the assumption that breakage would be exceedingly rare.[58] While Paraguay Educa was too small a project to broker a deal with Quanta directly, as Uruguay had, Plan Ceibal sold Paraguay Educa several batches of repair parts out of goodwill—though each time, this supply was quickly exhausted.

Thus, Paraguay Educa had to look to other solutions, many of which focused on individual or familial responsibility for keeping laptops in good repair. One initiative that began during my fieldwork in 2010 used parts from two broken laptops belonging to children in the same family (siblings or cousins) to make one working laptop that the children then co-owned. Around the same time, Paraguay Educa also embarked on a campaign publicizing how to care for XO laptops, detailing not only under what conditions the laptops should be used (sitting down, with the laptop resting on a solid surface and plugged in if possible) but how to clean them and do simple repairs. Paraguay Educa employees showed videos on the topic to students, and some teachers made posters about how to care for XO laptops to hang in their classrooms, emphasizing not the ruggedness of the laptops but their fragility and the expense of replacement parts.

Even with these changes, the situation had worsened considerably by November 2013, when I conducted follow-up fieldwork. The NGO still had not been able to secure a steady supply of parts and now lacked the money to pay for them. As a consequence, I encountered hundreds of broken laptops stacked in Paraguay Educa's office in Caacupé and heard stories of many more broken machines that remained in the hands of their owners. One interviewee estimated that, generously counted, only 40% of laptops were still usable at all. "The rest are very broken," she explained. She had tried to start a program to have children share laptops, "but the kids did not want to lend their computers. They would say, 'Oh, did you break yours? I do not want to break mine,' and they were very possessive."

4. Coping with Breakdown

Coping with these problems—charging laptops, installing software, and dealing with breakage—in the classroom was a constant source of stress for teachers throughout the project. With no spare laptops in schools, aside from the teacher's own, teachers had to develop strategies for dealing with the common situation in which not every student had a working laptop. Almost every teacher I interviewed commented on the difficulty of dealing with this problem. One school's coordinator stated in an interview that the problem effectively doubled the work of those teachers who kept trying: "The biggest problem is broken machines; when 50% [of students] don't have machines, then you can't use them with that group, or *the teacher has to do double the work* to have things for both the children with XOs and the children with notebooks. For some, the notebook is more convenient and comfortable, especially with computer malfunctions."

The strategy of the fourth-grade teacher in the vignette above—to have students without laptops do the same assignment in their notebooks—was common and relatively easy to implement when the lesson plan allowed for it, though it also limited what could be taught, as we will see. I also observed classes where teachers would pair students who did not have laptops with those who did, which allowed those without laptops at least some laptop time. However, paired children were often given very little control of the laptop by its owner. In some classrooms I observed, including the fourth-grade classroom described above, paired children stopped paying attention and became disruptive instead.[59] A few teachers, most of them in one large school whose leadership and trainers championed the idea, had students borrow laptops from other family members in the school. This took time and did not always work; sometimes the other class was also using the XO, or sometimes the student did not have family members with working laptops in the school. When it did work, however, the lesson was much easier in a classroom that at least temporarily had one laptop per child. Faced with these difficulties, many teachers simply gave up; they never used XO laptops in their classes.

Teacher Training, *Formadores*, and the Labor of Translation

From the picture of laptop use and breakdown thus far in this chapter, it would seem that OLPC's charismatic potential failed to be translated at all among teachers in Paraguay. But this is not wholly accurate: during my fieldwork, I noted that some teachers did keep trying to use laptops in their classrooms, despite the labor it took to cope with the various kinds of infrastructural breakdown they encountered. The reason for this was that the laptop *was* charismatic for them—though, as we will see next, this charisma was different than it was for members of OLPC, Paraguay Educa, and the technology world. In this section, we will turn from physical infrastructure and maintenance to examine some of the social factors that aided the translation of OLPC's charisma from MIT's founders and Paraguay Educa to the schools and teachers of Caacupé.

Paraguay Educa fulfilled an early promise it had made to President Lugo to provide teacher training in order to encourage classroom use of the XO laptop.[60] Though OLPC often talked of disrupting or even bypassing schools in its learning model, I had been told by one Paraguay Educa employee that this teacher training was a condition, along with *cédula* registration, for getting access to the schools for infrastructural upgrades.

In December 2008, employees and a cadre of volunteers gave all teachers in the ten Phase I schools a week of training in how to use their new XO laptops. In these training sessions, Paraguay Educa had its work cut out for it. In a survey of teachers that I conducted in 2010, I found that before receiving an XO laptop, only one in four teachers had a computer at home, and only one in eight had internet access. In contrast, over half had no access to a computer. (Though low compared to the United States, this level of access among teachers was higher than the one in ten Paraguayans overall who had a computer at home in 2009.)[61] Although all had basic-featured mobile phones, and nearly all had televisions at home, experience with this technology did not translate to computer use.[62] Thus, there were a lot of basic functions that Paraguay Educa had to cover in this training session: what is the internet, and how does one access it? What kinds of websites might be useful for teaching, how do you find them, and how do you navigate them? How does the laptop's camera work? What about word processing? What is email, how does one sign up for it, and how does one check it?

Paraguay Educa's training session was able to cover a lot of this basic use. After that, based on the model of cultural change that OLPC promoted, Paraguay Educa expected that teachers would be so excited about the possibilities provided by the XO that they would want to learn more about their laptops. The machine would speak for itself, with no additional translation needed; its charismatic promises could take over and help teachers imagine new technological futures for their classrooms—futures that would no doubt be both seductive and easy to implement. Of course, OLPC leadership had made statements about teachers and schools that were generally not very positive—statements that Paraguay Educa was aware of as well. The NGO's employees, moreover, were well aware of OLPC's stories of children taking to computers like fish to water and of children teaching the clueless adults in their lives how to use the devices.[63] Still, Paraguay Educa hoped that just as it found the laptop charismatic, teachers would too. The NGO honored Lugo's wishes to provide training and had made sure that teachers would have the rest of the summer—December 2008 to February 2009—to explore the laptops and integrate them into their curriculum plans for the coming year.

But as the 2009 school year got underway in Caacupé, it was soon apparent that this charismatic vision was not locally legible to the teachers or the students involved with the project. Even teachers who would later become advocates for the project told me that although the December 2008 training session was helpful for learning the basics of laptop use, they were left with little idea how the laptop could be a tool for learning. As one explained, "We were trained by a technologist who knew a lot of what the internet is, but pedagogically we did not know how to use the XO." With no ongoing support and little insight into what benefits the laptops were supposed to provide, teachers simply did not use them during the 2009 school year, and few students used them either. Teachers described continuing to struggle with basic operations, such as searching the web and checking their email—much less exploring some of the more complicated constructionist programs on the machine. Some were put off by difficulties using a laptop scaled for children, with unfamiliar software and sometimes-unreliable hardware. One fifth-grade teacher later admitted to me, "When I did the teacher training, it was one week and only about the internet and Paint and Write. We did not know what to do later."

As a result, in the first year of the project, the meanings that teachers (and students, as we will see in the next chapter) developed about the XO laptop diverged significantly from OLPC's—and most told me that they did not use their laptops at all. Teachers told me that they viewed the XO laptop more as a brightly colored plastic plaything than as a useful tool for learning in that first year, a judgment that had lasting effects. One school director said that she and her colleagues initially thought the laptop was "just a little toy for games." And it was clear from her tone that its being a toy was equivalent to being written off entirely. In short, the laptop was not immediately charismatic in Caacupé, and this failure of OLPC's technologically determinist model of cultural change threatened to sink the whole project into another story of disuse and no impact.

Could Paraguay Educa translate the charisma it saw in the machine and salvage the project? The NGO—although busy with infrastructure, maintenance, and the ever-present necessity for fundraising—was not blind to this lack of use. Around July 2009, just a few months after students in Phase I received their laptops, the founders enlisted the help of an Asunción-based education specialist to consult on the project and, shortly after that, brought her on full time as the director of Paraguay Educa's newly created education division. One fifth-grade teacher recounted an August 2009 visit from this education director, during which both realized that Paraguay

Educa would need to take a more active role in shaping the uses and meanings of the XO laptop if the machine was to be seen as something more than a toy. The teacher described that at this first visit, the director "said in a strong tone, 'You're using the [paper] notebook [and not the laptop].'" The teacher continued, "I didn't know what to say. I only knew how to turn the laptop on and use the internet, because that's what I was taught, but I didn't know how to use it pedagogically. I told her I can't teach what I don't know."

The NGO stepped up to the challenge to shift the initial opinions that teachers and students formed about its project toward ones that better matched the charisma it saw in OLPC. As the 2009 school year wound to a close in November, Paraguay Educa decided that the project needed additional day-to-day support to bridge what it saw as the laptop's charismatic potential with the messy realities of classrooms and schoolyards in Caacupé—and it changed strategies. Acknowledging that the laptop by itself was not sparking the kinds of exploration that the NGO had hoped for, it raised more money to hire a set of full-time *formadores*, or trainers, to do this translation work alongside teachers in the schools.

Paraguay's *formadores* program followed the model that Plan Ceibal, Uruguay's government-run OLPC project, had established, wherein each school nationwide had a dedicated technology specialist to help teachers integrate the laptops into classroom activities and to address technical problems. Paraguay Educa had specific demands for its trainers: they had to be based in or near Caacupé, they had to have experience teaching as well as programming, and they had to be excited about OLPC's promises. To join Paraguay Educa in early 2010, the fifteen trainers whom the NGO hired put aside teaching elementary, high-school, or college classes; working technical-support or programming jobs; or studying for advanced degrees. When the 2010 school year started in late February, these trainers, guided and paid by Paraguay Educa, began working daily in the ten Phase I schools, serving as local mouthpieces for the Asunción-based NGO by promoting the idea of the XO as a learning device, giving concrete suggestions for lessons that incorporated the laptop, and—when possible—trying to convey the charismatic vision that motivated the project, such as inspiring children to explore themselves and unlocking their natural creativity. Paraguay Educa also hired three additional full-time staff members for the newly formed education division: two curriculum developers and one extracurricular-activities planner.

During the 2010 school year, Paraguay Educa saw this considerable investment start to pay off: laptop usage in classrooms and homes started to rise. I heard the same story, over and over, from directors and teachers at all the schools: having someone whose job it was to educate the school staff about the laptop and its learning theories transformed the project from one that was completely marginalized in the classroom and beyond to one that was at least somewhat more integrated—though still not as central as OLPC or Paraguay Educa might have liked. "Before we had a trainer, we thought the laptops were just a toy for games," one school director explained to me, but since "we've learned to use them with [our trainer], they are much more important now." A teacher at another school echoed this sentiment: "With a trainer in the school, we use the laptops much more than the year before, when I didn't think laptops would be useful in class because I still didn't know the possibilities." I heard similar stories again and again in interviews: "Last year, we didn't have any support staff for the XO, and we weren't able to learn about all of the activities in training. ... [Now] our in-school trainer helps a lot," a third teacher stated. Another asserted, "It's so much easier with the trainers. Last year, we didn't use the laptop at all; it was impossible without them [the trainers]."

As Paraguay Educa prepared for Phase II of the project—giving laptops to students in the remaining twenty-six schools in the area—it put into practice what it had learned in Phase I. The teachers from the twenty-six Phase II schools received significantly more, and more pedagogically oriented, training than their Phase I counterparts. They received their laptops ten months before their students did. They attended three weeks of full-time training with Paraguay Educa's teacher trainers—who by then had half a year of experience using the XO laptops pedagogically—over winter break in July 2010, much of which I observed. They also attended ongoing supplemental training every month after that. Moreover, many benefited from contact with colleagues at Phase I schools who had on-the-ground experience with the laptops.

Paraguay Educa, however, did not have the money to grow the trainer program to provide all of these Phase II schools with a full-time trainer on site. Instead, the fifteen original trainers, like the technical staff, began to rotate between schools after laptops were given to Phase II students in

May 2011, generally spending a couple of days a week at each school. To make up for this reduction in trainer time, Paraguay Educa tried to harness volunteer labor from some of the most enthusiastic teachers in the program, whom it distinguished as *maestros dinamizadores* (teacher facilitators). It provided these teachers with some leadership-training workshops and encouraged them to be proactive with fellow teachers at their schools, taking over some of the duties the trainers used to have.

Tripling its staff to add the teacher-trainer program was not cheap for Paraguay Educa, and the NGO was concerned both with sustaining the program and with figuring out how it could scale if the project were expanded. Its budget was limited, and like many NGOs, it had no guarantees of long-term funding. Indeed, as time wore on and Paraguay Educa was unable to raise enough money to cover the trainers' salaries, this ongoing expense became too much for the NGO, and it gradually laid off these employees between 2011 and 2014. Paraguay Educa hoped that by then the charismatic promise of the laptops would be well understood, their use well entrenched in classrooms, and volunteer teacher facilitators well positioned to help as needed. But other factors confounded this attempted transfer of power, as we will see next.

Constructionism versus the Internet: The Limits of Translation

Paraguay Educa's trainer program was clearly integral to increasing laptop use. In 2010, I witnessed the trainers providing teachers with ongoing encouragement to try new classroom arrangements and more child-centered learning models. They also encouraged conversations about pedagogical practices, what worked best, and what resources were available to help. But were they able to translate the laptop's charisma as well? Did the charismatic promises of the XO itself, and of the constructionist learning model it was supposed to embody, become legible to the teachers in Paraguay through this labor?

With the help of the teacher trainers, some of the teachers and school administrators in the project did seem to understand this charismatic vision, at least when it was presented as an ideal; they liked the vision of child-centered learning that the laptop symbolized. Did this translate into using constructionist activities such as Scratch, Turtle Blocks, or eToys in the classroom? These were the kinds of programs that OLPC's leadership had asserted would set the XO laptop apart, after all, and Paraguay Educa was championing their use. To find out, we can return to the description of the fourth-grade lesson begun above to see a fairly representative example of the kind of pedagogical role that the laptop played in the classroom.

Forty-five minutes had elapsed since the teacher I was observing that day had first asked students to take out their XO laptops. Only nine of the fourteen students in attendance had working laptops at the beginning, and only two had the program she wanted to use, Tux Paint, installed. Losing three more laptops to drained batteries in the process, the teacher had installed the software on four additional laptops and was finally ready to start her lesson. She wrote "Alimentos" (foods) at the top of a blank panel of the chalkboard and then "según su origen" (by origin) underneath. She asked, "What are some foods?" The class brainstormed together for a few minutes: first the categories of animals, vegetables, and minerals, and then examples of each (cow, lettuce, salt). The girls participated more; the two boys with working laptops intently stared at their screens and ignored the teacher. The one whose screen I could see had closed Tux Paint and had the browse activity open to search for "chistes" (jokes). Then, another laptop ran out of battery power.

On the teacher's instruction, students started drawing examples of foods on their laptops or in their paper notebooks. Those with laptops drew shaky lines with their trackpads to divide the categories, often holding down the keys at the four corners of their keyboards to reset the trackpad hardware when it got jittery or unresponsive. They then used Tux Paint's built-in image "stamps" to generate examples in the three categories. The cartoonish sounds of Tux Paint's stamps emanated out of the XO laptops' tinny speakers, ricocheting off the painted cinderblock walls of the room. Students without laptops wrote out examples and drew pictures in their notebooks instead. After about ten minutes of working, the teacher announced that the rest would be homework for tomorrow and asked children to put their laptops away and take out their notebooks to copy a poem in Guaraní from the board in the half hour or so before school ended. Most students complied, but two boys kept their laptops out, half-open, and peeked at them from time to time. I peeked at one of them myself and found that the student was downloading an episode of *Naruto*, a popular anime series.

...

This lesson illustrates some additional negotiations that teachers encountered in trying to use XO laptops in the classroom. In particular, none of the computer uses that I noted that day were particularly XO specific. Even though this teacher professed enthusiasm for the program and was identified as a teacher facilitator by Paraguay Educa, in her classroom that day the XO laptop did not differ much from a generic computer. Likewise, the activity Tux Paint—though developed for children, relatively easy to understand, and fun to use—was also not specifically constructionist. It was open source, but it did not encourage deep, embodied, passionate exploration any more than any other drawing program. In fact, drawing on the XO laptop was inferior to drawing on paper in several ways, even after the software was loaded and working. It was beset by the difficulty of using the unreliable trackpad rather than a pencil, the temptation to categorize only the available images instead of thinking of other items and drawing them freehand, and, of course, the ever-present temptation of the internet.

The relatively frequent use of Tux Paint that I witnessed across many classroom observations illustrates a broader issue of which activities were considered best to use in class. The topic was a point of frequent and sometimes-heated negotiations between Paraguay Educa and teachers and was a site where different meanings of the machine were particularly evident. The staff of Paraguay Educa, like the OLPC leadership, wanted students to become enamored with the constructionist activities on the laptop, especially Scratch and Turtle Art. But teachers consistently described these two activities, along with eToys, as the hardest to incorporate into the classroom. These constructionist activities all had similar interfaces in which command "blocks" were grouped to direct the computer's actions in what amounted to visual programming, and all became even more difficult to manage with a jumpy trackpad—much less in classrooms where not all students had working machines. Even in schools that encouraged teachers to use these activities, only a small fraction of teachers were willing to put in the considerable unpaid time needed to learn them. Overall, Scratch, Turtle Art, and eToys were rarely used in the classroom unless a trainer was teaching with them directly.

Moreover, some (though not all) of the lessons I did watch that incorporated these activities were demonstration lessons rather than hands-on lessons, in which the trainer played what amounted to a video that they had developed with the tools it provided. Two such demonstrations I saw during fieldwork showed click-activated animations of the solar system and the functions of the digestive tract. Although such demonstrations could draw attention to the capabilities of these activities for students, they also replaced constructionism's goal of student-led exploration with a decidedly instructionist lesson directed by the teacher or trainer.[64]

What about those few teachers and trainers who really tried to embrace student-led learning with constructionist activities? I found that they had an uphill battle in encouraging students to explore these activities. Even when they overcame the technical difficulties of classroom use, they found that students seemed to be perpetually distracted by music, games, and videos on the internet (a theme we will explore more in the next chapter). In fact, teachers and students both agreed that the web browser was the easiest activity to use and the most compelling to all involved; it was important enough to children that they never uninstalled it. Based on teacher reports and my own observations, teachers used this activity almost every time they used the laptop, both in class and personally. "The focus is on using the internet in class," one teacher admitted in an interview, explaining that "it's easiest. We need to work more to connect other activities with lesson plans."

The reliance on using the web browser was so great that one of the few teachers who had embraced the idea of the laptop as a constructionist tool (María, who we will get to know better as the mother of the Scratchero/a siblings in chapter 5) labeled it as one of the project's biggest problems, after breakage. Not only were teachers relying on the internet in the classroom, she said, but they were allowing themselves to be distracted by it at meetings, much as their students were distracted by anime and jokes in the classroom: "We had a meeting with Phase I teachers, and everyone had their XOs and was checking their email; everyone was looking at their screen and not paying attention. Then we turned off the internet, and the whole room closed their XOs and began to pay attention. Nobody cared that everyone was doing something else."

María further advocated for teachers in Phase II of the project, who were in training during my 2010 fieldwork, to learn about their XO laptops without the internet. "The internet should only be a support," she said. In her ideal world, her colleagues would "use the XO like a [paper] notebook that has only limited use of the internet, because very few see the internet's educational side."[65] However, her opinions were not widely shared. When teachers at two of the twenty-six Phase II schools learned that Paraguay Educa would not be able to deliver internet to their schools because a privately-owned hill blocked the WiMAX signal, they walked out of training and had to be cajoled into coming back. "What is the point of this program if we do not have internet?" one teacher quipped to me in an interview.

Thus, the internet was the project's most charismatic feature for many of the teachers involved. I heard from some teachers that being connected to the internet meant being connected to the world. It would help them and their students learn. One teacher stated, "For these kids, nothing will surprise them. Teachers can be intimidated even by cash registers, but the kids are prepared for the future, and with luck they'll study more and more and make our city evolve into something great." A parent expressed similar sentiments about the internet. "We'll have a digital city," she predicted. "These kids won't be shy. Citizenship itself will change: there will be more reflection, and votes will really reflect the best candidate. They won't believe everything they're told because they can investigate anytime and anywhere." To these participants, what was most charismatic was not constructionism but the internet, and the kinds of predictions they made about its effects bore some similarity to hopes circulating more broadly in the early days of the World Wide Web.[66]

The other most popular activities on the XO were likewise not the constructionist ones, according to my observations and the aggregated usage statistics I collected with Paraguay Educa in fall 2010. After the web browser, the next most used activities were the word processor and the two drawing programs, all of which allowed teachers to straightforwardly replace parts of blackboard-and-paper lessons. Teachers were clear in their reasons for making only the easiest substitutions: it allowed the one-quarter of students whose laptops were broken to use their paper notebooks instead. Many teachers told me that it was enough work to incorporate the laptops into their lessons at all, much less to use the more complicated activities. One trainer echoed these findings in an interview. "Right now, both the teacher and the student use the easiest activities," he quipped.

Moreover, this trainer found that many teachers would only use the laptop if a trainer was actually present to help them. "In many cases, if there is no trainer, there is no use of the XO," he explained. Paraguay Educa's trainers were instructed to counter this tendency by encouraging teachers to develop lessons that incorporated constructionist activities and to fall back on the trainers only for guidance and support. Some trainers did make headway with a handful of teachers, most of whom they also nominated for the teacher-facilitator program. Other teachers wanted to enlist the trainers to just teach XO lessons directly to students, though, and some trainers were not very good at insisting that they do otherwise. One teacher explained this dynamic in her school, which was large enough to have two trainers:

> One trainer was supporting us better than the other. [Trainer 1] will just teach the classes, so the teachers depend on him and don't learn the skills themselves. When he's not there, the teachers don't do anything. On the other hand, [trainer 2] will see how you work, and you can talk with her; she's more open. She'll explain and after help you. I realized that this was what the role of the trainers should be. ... It's bad enough that a teacher will say that since today isn't the trainer day, we didn't use the XO today.

Many teachers found that it was considerable work to incorporate the laptops into their lessons at all, even with a trainer's help. The fourth-grade teacher in the vignette above, for instance, set out to teach a lesson on the XO on her own, even if it was not constructionist, without relying on the trainer. This level of active engagement with the laptop distinguished her from many of her peers. In contrast, the teacher in a second classroom that I briefly visited at the end of the morning session was not interested in using the laptop herself and called instead on the trainer to teach with the laptop while she graded assignments at her desk.

This disparity illustrates a persistent problem with some teachers: though nearly all teachers expressed admiration for the program in the abstract, the trainers and my own observations counted at least one-quarter of the teachers who never used the laptops in class unless a trainer taught for them.[67] My brief observations in Uruguay suggested similar results there: based on four full days of observations at four different schools, it seemed that Uruguay struggled to adequately train its large full-time support staff, and I was told that laptops were largely used only when the support staff was present to help.[68] From 2011 to 2013, Paraguay's trainer program was gradually cut back, and any classroom use that depended on a trainer's presence also subsided—despite the efforts of the teacher facilitators, as we will see next.

Paraguay Educa was thus not very successful in translating OLPC's charismatic promises for the XO—of the laptop as a tool for constructionist learning—into the classroom. Most teachers struggled to learn how to use the more constructionist activities, with some giving up on laptop use more generally and some handing over their classroom use to trainers. In contrast, the most used activities in the classroom were those that were easiest to understand and integrate into existing lesson plans, and most popular by far was the web browser. The XO was thus charismatic to teachers not because it could unlock the excitement of programming and programmatic thinking for students, but because it was a portal to the internet.

These kinds of different meanings arose in part from teachers' varied responses to the many demands on their time and loyalties—responses that OLPC's individualistic approach did not prepare local projects to recognize and handle. In addition to pressure from Paraguay Educa to use the laptops or specific software in the classroom, these demands included commands from the school director or coordinator, curricular requirements from the MEC, the time constraints of the school day, activism against teacher exploitation and low pay, and family or other personal concerns. Our look into a fourth-grade classroom showed that balancing the demands of the MEC's national curriculum against a school day that was only four hours long, *including* recess, could be particularly difficult. Another teacher spelled out the constraints in an interview, ultimately concluding that the laptops just took too much time to use in the classroom:

> We don't have a lot of time to use the XOs in class. I'd like to use them more—there are so many things to discover and do with it. But the ministry requires us to complete four lessons a day, including math, communication, and natural sciences, and you need at least forty-five minutes for the XO. These aren't activities they'll figure out in twenty minutes, and it takes time to open them up, to type—the children are slow.

Teachers were thus required to cover a lot of ground every day in class—challenges belied by their low pay. In 2010, teachers told me that they were paid 600,000 guaranies (about $120) a month for planning, teaching, and grading one four-hour school session five days a week. As public-sector workers, teachers were not protected by Paraguay's minimum-wage laws. They also received no benefits. These were points of frequent contention between the two national teachers' unions and the federal government.[69] In fact, during my fieldwork in 2010, the two largest teachers' unions would strike on alternate Thursdays for improved wages. Because so many teachers in the public schools were part of one of these two unions, schools often just cancelled classes completely when teachers were striking rather than attempt to operate with half of their staff.[70] On the other hand, Paraguay Educa's nongovernmental status was useful in garnering support: even though the NGO had to cooperate with the government, teachers saw it as separate and thus not competing for the same resources or complicit in their poor treatment.

To make ends meet, most teachers taught both morning and afternoon sessions. At larger schools, they would often teach a different set of students at the same grade level, but at smaller schools, teachers might teach a different grade level in the afternoon or even go to a different school to teach. Thus, whereas the school day was four hours long including half an hour of recess from a student's perspective, from a teacher's point of view the school day lasted from approximately six in the morning to after five in the evening. In addition to this, some teachers took another part-time job as well, such as teaching an evening class or helping a family member with their business. Though some did all of their grading during recesses and the last fifteen or thirty minutes of class and reused materials from previous years, others told me about spending Saturdays and Sundays grading lessons from the past week and creating lessons for the next week. Teachers were also required by the MEC to complete a number of unpaid hours of supplemental training every year. (Paraguay Educa had made an agreement with the government that training sessions for the XO laptop counted toward this requirement.) In some schools, teachers might be called on by the school administration to do other volunteer work for the school community, such as repainting the schools or teaching after-school programs. In the evenings, many focused on their families, and some were single mothers. All of these demands on their time left little extra for learning to use their laptops or integrating them into lessons.

The leadership at two of the ten schools in Phase I tried to align at least some of these pressures. One school coordinator who supported Paraguay Educa's mission helped the teachers under her supervision to develop curricula that used the XO in constructionist ways. She taught them how to upload a lesson to the school server and have children download it and complete it without needing to copy anything from the board (though students without laptops might still have to). She found, however, that few teachers would invest the considerable time she requested of them—four unpaid training sessions every week—to learn the laptop's constructionist activities. In an interview, she explained the range of responses these teachers had to the project. Some did not have the time, she said, whereas others joined schools after the training sessions and did not have a laptop. "[My daughter's] teacher had her baby," she explained in one case, "and another teacher [without a laptop] came to replace her."

Similarly, the director of another school, who had a longstanding interest in a variety of educational reforms, revised her school's mission around constructionist teaching with the XO and encouraged the teachers in her school to volunteer their time for curriculum development and support for fellow teachers, students, and students' parents. Even under her commanding authority, however, some teachers were less committed to the project. She explained to me, "I can see that we're innovating our old pedagogy with the XO. It's a difficult change. ... If the teacher has training, the XO, and space but no change of attitude, nothing happens." Her framing of the problem as one of individual "attitude" placed the blame on the teachers themselves rather than on structural issues such as a lack of time and pay for these extra duties.

The other eight schools in Phase I of the program in 2010 focused less on promoting constructionist learning in the face of other considerations. The fourth-grade teacher whose classroom was described earlier in this chapter did find the XO laptop charismatic, at least in a general way. However, her school's leadership was not so committed, and the trainers at her school were not very forceful, making her an anomaly who lacked the influence to rally her colleagues. Similarly, the coordinator at yet another school wanted to promote the project, but both she and the school's trainers lacked the leadership to recruit more teachers to use the XO in the face of an indifferent director. Paraguay Educa found that even though the practices of the most committed schools could inform strategies for reaching the others, it was very difficult to take into consideration the specific politics and practices already present in each school, much less shift them all to align with the project's goals. Moreover, in all schools, teachers chafed against the expectation that they put in many hours of unpaid labor to learn how to use the laptops. Overall, the expectation that teachers should engage with the program, and the shaming of those who did not, neglected to consider the many demands that teachers had on their time and the low pay they received for their work.

Charisma's Brittle Promises

At the United Nations Social Innovation Summit, the Open Mobile Summit, and several other talks in 2011, Nicholas Negroponte said that computers could be literally dropped out of helicopters, and children would be able to teach themselves to read with them. "When I say ... drop out of the helicopters, I mean it," he asserted. "It's like a Coke bottle falling out of the sky."[71] With statements like this, OLPC's leadership clearly wanted to be disruptive, and creating friction in schools was part of that disruption. Was OLPC able to achieve the kind of disruption it wanted?

Peru's OLPC project—which was the largest in the world before Uruguay surpassed it in 2014—more closely followed Negroponte's suggestion to just give out laptops and walk away.[72] Two successive governments distributed over one million laptops in total—the first focused on rural areas to which the administration really did have to deliver the machines to children and teachers via helicopter—but did not follow up with much social or technical support.[73] Most teachers received either a week of training or none at all, few schools had internet connections, and some rural schools did not even have power to charge the laptops.[74] A 2012 analysis by the Inter-American Development Bank showed little change among the more than five hundred thousand Peruvian children who received XO laptops in 2010 or earlier.[75] This critique echoed more broadly what observers on the ground reported anecdotally and what I witnessed myself during a two-week visit to Peru in September 2010: the project was underresourced from the beginning and never had the infrastructural investments of power or internet access that were promised, nor did it have adequate social investments in teacher training, maintenance, or other support.[76] As a result, many of the laptops were rarely used—in some cases they were locked in classroom cupboards—so it is no wonder that they had little effect, disruptive or otherwise.[77]

Uruguay's model, in contrast, was closer to Paraguay's, as we have seen; it enjoyed more intensive infrastructural and social support. Plan Ceibal's stated goal was to eliminate the digital divide nationwide, which the laptop program accomplished almost by definition: all children have received computers, and all schools have installed Wi-Fi, which means that children across the country now live within one hundred meters of a free internet connection. But what of the laptops' disruptive potential—to be a "Trojan horse" for constructionist learning? A pair of 2009 surveys indicated that the laptops were popular with all involved but admitted that the project had not produced the groundbreaking change that OLPC had envisioned.[78] Moreover, one survey estimated that the number of inoperably broken laptops in Uruguay in 2009—less than two years after the project's first nationwide laptop distribution had started and only barely after it had finished—was between 25% and 35%, despite Plan Ceibal's extensive investment in repair facilities. In November 2013, the national university in Uruguay published a working paper on the educational impacts of Plan Ceibal, finding no gains in reading or mathematics.[79]

Some researchers in education have joined OLPC contributors in discounting the results of such surveys and standardized tests as flawed or, at best, not measuring the right thing. This chapter contextualizes such results (of which Appendix A contributes another) with an ethnographic account of the translation work that took place in Paraguay's OLPC project, highlighting similarities and differences with Uruguay's and Peru's programs along the way. We have explored how OLPC's hopes for disruption played out in Paraguay, considered by some in the OLPC community to be the most successful project—and a project that, like Uruguay's but unlike Peru's, put in extensive infrastructure and social labor to try to make the project succeed. Given, however, that the constructionist activities that OLPC championed got little use—especially without Paraguay Educa's trainers directly involved—and that the laptop's very design caused enough problems and delays in the classroom that many teachers gave up in frustration, it would seem that the project was not the success that OLPC or Paraguay Educa wanted. Like his comparison of XO laptops to Trojan horses, Negroponte's "Coke bottle" reference to the 1980 South African movie *The Gods Must Be Crazy* neglected to consider the impact of the empty Coke bottle that was carelessly dropped from a plane on the San tribe depicted in the film: chaos, strife, and ultimately rejection.[80]

Paraguay Educa—whose members came from the cosmopolitan upper crust of Paraguay and who were familiar with the social imaginaries important to OLPC—found OLPC's vision of the laptop as a charismatic tool for unlocking children's natural curiosity not only legible but inspiring and

personally meaningful. But this chapter shows that this charisma was subjective, not universal. OLPC's vision needed considerable ongoing work to
be translated into the day-to-day realities of Caacupé, including infrastructural investments in schools, maintenance and repair work, and full-time
teacher trainers. Even then, the translation was incomplete; it failed to
account for the structural limitations of Caacupé's schools and the social
worlds of its teachers. When it came up against the messiness of day-to-
day life, OLPC's charisma thus became brittle. The cultural change that
One Laptop per Child and Paraguay Educa sought was neither effortless
nor inevitable; it was instead prone to breakdown and in constant need
of repair.

4 Little Toys, Media Machines, and the Limits of Charisma

Kids that I know, for instance in that picture, sleep with their laptop. I mean, they're not going to let these break, and Mary Lou Jepsen's done a heck of a job in making them very repairable, so that 95% of the maintenance is done by the kids, actually.

—Nicholas Negroponte, "OLPC Analyst Meeting"

This chapter examines the kind of laptop use that OLPC leadership claimed would make the biggest difference in children's learning: use that was initiated and directed by children themselves, not by a teacher for an assignment. This out-of-classroom learning was the cornerstone of the project, just as it was the cornerstone of previous constructionist projects. Even if classroom XO use was fraught, the laptop's subversive qualities were meant to be charismatic to children in particular, not necessarily their teachers— the children would soon leapfrog past them anyway. This was the reason that OLPC's first core principle stated that children should own their own laptops, rather than schools, families, or the government. That way, they could have unfettered access to the machines, learning as much and as deeply as they wanted. And, OLPC reasoned, of course children would want to explore. Seymour Papert considered the computer to be "the Proteus of machines," as he explained in the prologue of his best-selling book *Mindstorms* (1980). "Because it can take on a thousand forms and can serve a thousand functions," Papert explained, a computer "can appeal to a thousand tastes."[1]

This desire to support out-of-classroom learning motivated the design of the laptop and much of the content included on it. As we saw in chapter 2, the XO was more rugged than most laptops, with no internal moving parts, minimal connectors to break or get dirty, a solid silicone keyboard to protect against water and dust, and a rubber and plastic case that could help absorb the impact of being dropped. OLPC's leaders, moreover, boasted that its parts were easy for children themselves to repair or replace because the laptop was designed to encourage children to tinker, both with the hardware—accessible via standardized screws, with extras stored in the handle—and the open-source software.[2] This would let them delve as deeply as they wanted into the workings of the machine, the way that OLPC's contributors themselves had done with computers in their youth and in the hands-on way that constructionist learning encourages. However, the need for repair would be minimal, according to OLPC leadership. In a 2005 interview, Papert confidently asserted, "If kids value the computers, they'll take care of them."[3]

What did child-directed laptop use in Paraguay look like, particularly after children's initial excitement for owning a laptop wore off? Were the children in Paraguay captivated by the charismatic promises that the machine made: to unlock their natural creativity and learn with the computer, to foster a love for mathematics and programming, and to use their newfound passions and skills to transform their countries? Were these promises fulfilled? By the time of my fieldwork in the second half of 2010, children had owned their XO laptops for almost a year and a half, plenty of time for the laptops' novelty to be subsumed by more regular usage patterns. Across six months of fieldwork focused on the ten Phase I schools that had laptops in 2010, I observed children's undirected play during recess and before and after school, as well as in the homes of dozens of families I interviewed in and around the town of Caacupé. In these interviews, I talked to children about their typical laptop use and asked them to show me their XO use during the past few weeks via the Journal activity ("activity" was OLPC's term for a program), which recorded which other activities had been opened and any work done in them. I corroborated these observations with a look at the aggregated statistics of the entries in the Journal activity of all four thousand participants in the program in 2010, which I helped compile in October of that year.

What I found was striking: about two-third of children hardly used their laptops at all. This proportion was consistent across my field observations of children's leisure time at school and home, as well as in interviews with a representative sample of children in the program. Fifteen percent of children did not use their laptops because they were unusably broken, but the rest—about half of all children—were just not very interested in using the XO or found using the laptop more frustrating than engaging. The remaining one-third were using their XO laptops—but for purposes for which its hardware was ill suited: media consumption. Captivated primarily by the internet, not by the unique features of the XO, these children learned and then taught one another to overcome the limitations of the machine to play video games, music, videos, and more, passing files to one another. Children and teachers both said that the part of the program that was most captivating for them was the internet.

These findings raise several questions. How could a device designed to be "the Proteus of machines" fail to engage half of its target population? When the device was designed to be so rugged and easy to fix that a child was supposed to be able to repair it, how could 15% of laptops be unusably broken not even a year and a half after Paraguay Educa had distributed them? Finally, why were nearly all of the remaining children captivated not by the constructionist programs on the XO laptop but by media on the internet?

This chapter explores the limits of the XO laptop's charisma among Paraguayan children. This in turn highlights the limitations of OLPC's nostalgic design, particularly the disconnect between these children and the yearner child that Papert had described. This "yearner," Papert's name for the kind of child who evoked the social imaginary of the technically precocious boy, also seemed to influence OLPC developers' memories of their own childhood experiences and even their identities. They moreover implied that this imaginary could be universal and was above cultural influences—even when many readily acknowledged that they were exceptions among their peers. Instead, the limits and transformations of charisma that we will explore here highlight the agency of the children in Paraguay, who already had rich, full lives without a laptop—especially a laptop they found unrewarding to use—and most of whom were embedded in worlds in which advertising and media corporations held considerable sway. By closely attending to the realities of day-to-day use and the meanings developed on the ground, this chapter contextualizes OLPC's charisma; it accounts for children's agency in deciding how to use their laptops, on the one hand, and how that agency is circumscribed and shaped, on the other.

"Little Toys": When Most Use Is Nonuse

One fourth-grade boy, whom I will call Roberto and whose mother was one of the Paraguay Educa trainers promoting XO laptop use in the classroom, told me in no uncertain terms why he was not interested in his laptop. He said he would rather play soccer, watch television, play games on his PlayStation, or surf the internet via his mother's laptop instead of using his XO at home.[4] "I don't use my XO much; I don't like it. Sometimes I play sports games, but that's all," Roberto reported to me, showing me the last entry in his Journal program: a soccer game that had last been opened several months prior. He followed up with a story about how one Paraguay Educa employee "told me some kids think it's just a toy and nothing important."

Other children also told me that they thought the laptop was a "little toy" (*juegito*) rather than a machine for learning. Most, like Roberto, simply had little interest in what the laptop had to offer. These children instead spent their time, whether by choice or necessity, on other activities: helping babysit siblings or manage the family business, playing soccer, watching television (*The Simpsons* was particularly popular during my 2010 fieldwork), or spending time with family or friends rather than using the XO laptop. This lack of interest was especially common among households that had another computer, around 15% of the households I visited.[5] Children familiar with other computers in particular often complained that their XO was much slower, its memory was too limited, it was harder to use, and it could not connect to the internet at home, which was almost always via a USB stick that connected to the cellular network.[6]

Roberto's mother hypothesized that having an XO laptop "probably has a bigger effect on children who don't have access to other technologies, only their XO." After all, Roberto's XO competed for attention with a television, a PlayStation, and his mother's computer. However, even children whose sole access to computers was via the XO laptop echoed the same sentiments. One fifth-grade girl at a larger public school explained that she just didn't have time for the XO because her day was already too full: "In the afternoons I play, and I help my mom take care of my brother. I take care of him Saturdays too. In the morning I go to school." And it was rare for children to not have access to any other media technologies. The vast majority of the households that I visited had televisions, a proportion that has been relatively constant in Paraguay since 1996.[7] Most of those who did not have televisions lived in more rural areas and were very poor, but even so, televisions were common in the rural areas I visited. In fact, many children and their parents reported knowing about computers and the internet—including the perceived dangers of both (especially the latter)—from television programs.[8] Radios were even more ubiquitous, and quite a few households and shops had music—and advertisements—playing in the background all day. Thus, even if most XO laptops were not competing for attention with other computers at home, they were competing with television and other media technologies.

Furthermore, whether children had access to another computer or not, or to other technologies or not, most told me that they were frustrated by certain aspects of the XO laptop. When I asked them in interviews if there was something they would like to change about the laptop, nearly all children mentioned waiting too long for the laptop to boot or for activities to start, struggling to use the jumpy and unreliable trackpad, quickly filling up the one-gigabyte solid-state hard drive, and running out of power. As one mother explained, "It takes three minutes for the computer to turn on. Opening the computer takes all that, and just when it is turned on, its battery is finished." Though their XO laptops used less power than a typical laptop, children also complained about the XO laptop's ever-shortening battery life as the batteries that came with their laptops in 2009 gradually wore out. Finding ways to charge the batteries with broken chargers and a paucity of power outlets was another source of frustration. These problems were annoying enough for some children that they just chose not to use their laptops rather than to put up with them.

One social factor that affected some children's interest in their XO laptops was their teachers' level of interest. Roberto was one of several children I interviewed who cited this as a reason, even though his mother was working for Paraguay Educa: "The teacher influences a lot because I used it more last year with my old teacher," he said. The mother of a second-grade student at the same school who also had a computer at home explained her son's lack of laptop use the same way, excusing his teachers for having many other things to balance but still wishing laptop usage were higher:

> The children only use the XO with the teacher trainers [employed by Paraguay Educa]; it's not integrated with teachers' lessons. At [my son's school], they have so many special classes [with different teachers], so it's difficult to develop. It

might be better if there was a specific class for the XO. They use the laptop twice a week now at most. I think they don't use it more because it would make them fall behind in other subjects. That is what his teacher told me. There are many things that together mean that the laptop isn't used very much.

Nevertheless, I did notice that even in classrooms where teachers worked hard to use the XO laptop, up to half of the students still had little interest in the laptop except for required schoolwork. At a school that had developed the most ambitious goals for integrating XO laptops into the curriculum, one child reported using the laptop in school and for homework but little else: "At school I mostly use the internet. We use Browse [Navegar, the web browser] and Write [Escribir, the word processor]. At home I don't use it as much—just for homework and sometimes games." Similarly, having family members or friends who engaged with the laptop inspired a few children to use them more, but others who did not care for using the XO simply lent their laptops to peers who were more interested but were not in the program or had trouble with their own laptops.

Based on reports from children, parents, teachers, and teacher trainers, the perception that the laptops were little toys developed toward the beginning of the program, based on the appearance of the laptops and the lack of information that many families had about the program—as the previous chapter explored in more detail. It may seem at first blush that the perception of the laptops as toys might actually match the intended uses of the laptop by Papert, Negroponte, and the rest of OLPC. After all, they did want the laptop to be playful. Still, there is a big difference between using the laptop in a playful, creative way, as OLPC has described, and not using it at all because it appeared to be a toy. In this way, seeing the laptop as a little toy was a misreading of the laptop's focus on play—one that, we will see next, had real material consequences.

Whereas a couple of schools, with Paraguay Educa's help, eventually organized workshops to show parents the basic functions and inform them of the educational potential of the XO laptops, most schools did not, and most parents at the schools that did have workshops could not attend them. Instead, in keeping with OLPC's early claims that undirected child use was most important, most children developed their own meanings around and uses of the laptop in that first year of the program. For a majority of those children, the laptop was simply uninteresting as a tool for independent exploration.

These children's disinterest created a pattern of nonuse in Paraguay Educa's One Laptop per Child program, showing that the XO laptop's charisma had clear limits: for about half of children, the laptops appeared not to resonate at all. This challenges one of the claims in Papert's writings on constructionist learning that paints those who don't engage as "schoolers," lost souls already conditioned into a life of acquiescence to authority. What we find when we examine why these children did not engage with computers is not an uninteresting mass of automaton "schoolers" but a set of diverse children with their own opinions, motivations, and lives—ones that may not include an often frustrating-to-use laptop.

This group of nonusers also challenges Papert's assertion of computers as the "Proteus of machines," with something that could appeal to everyone.[9] The rejection of the XO laptop by half of the program's child-beneficiaries challenges the notion that computers—and especially these specific computers, designed as they were to elicit programming from children—are somehow radically different, and far more compelling, than previous technological innovations, making it safe to ignore the tepid (generally incremental at best) results of the introduction of previous technologies touted as solutions or replacements for traditional classroom education. Although it is true that computers have an interactivity that broadcast technologies such as televisions and radios lack, the ability to broadcast information that one-to-one technologies such as telephones lack, and a degree of customizability that leaves many previous technologies in the dust, the cultural revolutions promised by them are not so different from the cultural revolutions promised by previous technologies. Perhaps it is safe to say at this point, then, that computers may be useful for education but fail to deliver the promised complete overhaul—and that OLPC's XO laptops, which Papert proclaimed an exception even to other computer interventions, are in reality no different.

Unruly Machines: Limits to Repairability

Although just over half of children were just not very interested in using their XO laptops on their own, about one in six children didn't use their laptops during my fieldwork in 2010 because their machines were broken—a proportion I drew directly from Paraguay Educa's database of breakage reports in August 2010 and then corroborated in my classroom and schoolyard observations. Some of these children were initially laptop users, but breakage forced them to be nonusers, aside from a lucky few who were able to get repair parts or borrow a laptop from a sibling or cousin.

Some teachers, teacher trainers, and directors said that allowing children to take laptops home led to many more breakages, because most breakages happened at home. The number of children who took care of themselves after school and roughhoused with siblings during my visits to their houses would suggest that this may well have been the case. One second-grade teacher bemoaned this policy and highlighted the problems these broken machines caused in the classroom: "It'd be better if we left the laptops at school. So many are broken on the way home; we've had so many accidents, so many broken screens. Some stay at school late [to use the school internet]. It'd be better to just leave laptops here. ... We suffer in class when many children don't have machines. They have to use it with a partner, which neither likes." One teacher trainer at another school blamed parents for this breakage: "Laptops—especially screens—get broken mostly at home, and parents don't care enough to make their kids take care of the technology. They stumble and fall, and then it's broken. It's a failure for Phase I. For example, don't just pull the cable to unplug it [*mimes yanking the cord instead of detaching the transformer at the end*] and play with the cable like this [*mimes spinning the cable*]. It stretches, and obviously it'll break." More frequently, however, parents worked late, and many seemed to be unaware of how expensive the laptops were to repair or did not see enough value in the machines to encourage their children to protect them.

If laptop breakage did happen during unsupervised time at home (or on the way home), then those children who tended to have more unsupervised time—children in rural areas—might exhibit more breakage. This is just what the data showed. According to an August 2010 snapshot of Paraguay Educa's Inventario (inventory) system, fully 50% more rural laptops were broken than urban laptops, though reports of software problems were roughly equal between the two groups.[10] Even though I was not able to definitively answer what was behind this difference, I did observe that rural children generally had longer walks home and often spent more time outdoors with their laptops than urban children, meaning that the laptops were more often used while standing or walking where they could be dropped or caked with dust more easily. As they admitted in interviews, the parents of rural children tended to be less familiar with digital technologies; almost none had computers at home. They often worked long hours, leaving their children alone or in the care of siblings or other family members before and after school. All of these factors, taken together, may explain the increased breakage rates among rural children. This raises troubling implications for equity: rural children also tended to be poorer, so this trend reinforced some of the very socioeconomic divisions that the program was meant to alleviate.

Even when repair parts were available, most were prohibitively expensive for many Paraguayan families, who, during my fieldwork in 2010, were responsible for purchasing repair parts themselves. A replacement screen, for example, cost a family 303,600 guaranies (about $65), which for most families was a sizable chunk of their monthly income. Rural families, many of them subsistence farmers with side businesses selling homegrown or homemade products on the side of the road, were particularly unlikely to be able to afford repair parts, even though the repair labor was paid for by Paraguay Educa. As a result, aside from chargers (which, with the possibility of splicing, had roughly equal rates of repair in urban and rural schools), most hardware problems were repaired roughly twice as often in urban schools as in rural schools (see table 4.1). Thus, though the laptop program did in some ways help lessen the socioeconomic urban-rural divide in Caacupé, Paraguay, the divide persisted in both breakage and repair rates. Between 2010 and 2013, Paraguay Educa started subsidizing the cost of parts to help poorer families (though parts remained scarce and hard to procure).

Breakages were also gendered, following Paraguayan gender norms that, as in the United States, socialized boys into being more rambunctious than girls. Whereas software problems were roughly equal, my August 2010 data sample showed that boys had just over 30% more hardware problems than girls, especially the kinds of hardware problems that resulted from rough handling: broken screens (327 boys to 194 girls), chargers (456 boys to 398 girls), motherboards (30 boys to 11 girls), and hard drives (13 boys to 4 girls). While this could indicate that more boys engaged with their laptops in the first place, it meant that over time, fewer boys as well as fewer rural children had working laptops to use in the classroom or at home.

Overall, the relatively high incidence of laptop breakage disrupted the emerging usage patterns of some initially active laptop users and partially

Table 4.1.
Percentage of repairs in rural versus urban schools, by repair type

Problems fixed (%)	Rural	Urban	Urban:rural	Cost of part
Charger/battery	43%	45%	1.05	30,500 ₲ ($6.50)/ 71,000 ₲ ($15)
Keyboard/mouse	14%	27%	1.96	41,300 ₲ ($8.80)/ 66,800 ₲ ($14)
Screen	13%	28%	2.17	303,600 ₲ ($65)
Other hardware	14%	24%	1.76	(varied)
Software/OS	98%	99%	1.01	—
Activation	88%	99%	1.12	—
Average	59%	67%	1.13	—

Note that repair rates for keyboard/mouse, screen, and other hardware problems were twice as high in urban schools as in rural schools.

erased the leveling effect that the laptop had across socioeconomic classes. Although some children with broken laptops were able to borrow a laptop from a family member or (less commonly) a friend, their use of the borrowed machine was usually more circumscribed than use on their own laptops would be, as the lender and borrower sought to balance memory usage, program installation or uninstallation, and general care for the still-working machine. Because the lender sometimes had to reclaim their machine at least for classroom use and assignments, the borrower sometimes lacked the complete freedom with the borrowed machine that they would have had with their own machine.

In this way, the machine itself was "unruly": its breakage destabilized the meanings that OLPC and Paraguay Educa hoped to attach to it, as well as some of the meanings that some children themselves had started to develop. Moreover, the machines were unruly along lines that both were gendered and reinforced existing socioeconomic divides. This illustrates how privilege—through infrastructure, environment, and social support— can influence a child's opportunities in ways that may be largely invisible on an individual level but can be sizable in aggregate. Without explicitly attending to disparities in privilege, projects such as OLPC's not only risk recreating them but tend to do so.

Media Machines: Overcoming the XO Laptop's Limits

What of the one-third of children we have not yet discussed: the ones who were using their laptops and had not broken them? To open the analysis of their use, I will turn to a vignette from my field observations. When I arrived at the Phase I school on foot, thirty minutes before the morning session of classes started on a chilly day in August 2010, students in white shirts and blue pants and pinafores were trickling into the schoolyard alongside me, chatting and roughhousing with their friends. Some boys played a pickup *fútbol* game in the central courtyard. One girl swept out a classroom and the open-air hallway in front of it. Other children simply ran around the yard pell-mell or stood in small clusters in classroom doorways or hallways facing the courtyard, watching peers play and talking with friends. A few lingered around the school cantina in the corner of the courtyard, buying candy or cups of hot, sweet, milky tea to combat the still-chilly winter air. A couple of girls recognized me and ran up for a quick hug, a common expression of affection that many Paraguayan adults, especially teachers (which I was considered by many students, who did not fully understand "researcher" or "ethnographer"), gave freely to children.

Around one dozen small clusters of children, maybe one-quarter of those in the schoolyard that morning, crouched around sticker-covered XO laptops, giggling and hiding their screens when any adult approached, even though the teachers who occasionally passed by never tried to see what they were doing. When I asked them what they were up to on their laptops, some of these groups replied "Nada" (nothing) and hid their screens, but most let me see. Four small groups of boys were playing or watching *juegitos* (little games): two played a side-scrolling game called Vascolet, one played a game called Wear the Shirt (a *fútbol* game), and one played Super Mario Bros. in Wine, an open-source program loader that allowed users to run Windows software on Linux machines, which had been ported to the XO laptop.[11] This game playing was quite social: for each boy using a laptop, several more clustered around, their steaming breath mingling, to watch his screen. Another boy sitting by himself searched for "juegos con motos" (games with motorcycles) on the XO Planet website. A girl nearby searched for hip-hop music, and two boys searched for *reguetón* music ("reggaeton" in English) to play on the tinny XO speakers in the classroom or on the walk home from school. *Reguetón*'s characteristic *cha, ch-ch-cha* backbeat emanated from a few XO laptops around the courtyard, adding to the din of children's voices at play.

This vignette shows that the laptops were indeed popular among some children. It also illustrates what most unsupervised XO use looked like. The one-third of children who used their laptops fairly regularly were especially interested in finding ways to use the laptop as a platform for watching videos, playing video games, listening to music, and consuming other media. One question often asked about OLPC is whether these uses count as learning. More to the point, do they fit within the charismatic ideals of the project? The children I saw that chilly morning in August 2010—and many other mornings, afternoons, and evenings of fieldwork—did seem to match promotional pictures of children hunched over their XO laptops, pictures that OLPC and various individual OLPC projects (including Paraguay's) used as evidence that the project's promises were being fulfilled.[12] A closer examination of just what the children were doing on their laptops, however, brings up more complex questions of just what we mean by "learning" and whether what these children were doing counts.

To begin to answer these questions, we can return to the vignette above, which shows that the day-to-day use of OLPC's XO laptops in Paraguay did not completely coincide with OLPC's charismatic vision: nearly all of the children who used their laptops at all had no interest in the constructionist tools that were meant to help them intuitively and naturally learn mathematical thinking. The social imaginary of the technically precocious boy who found pleasure in technical mastery, which was behind the constructionist learning theory that motivated the laptop's design, did not seem to be something that resonated with them. Instead, the XO was, for the most part, another machine with which to consume media. Nearly all voluntary (non-teacher-directed, non-homework-related) XO use that I witnessed— whether before and after school, during recess, at home, and even during class—focused on game playing, video watching, music listening, and other media consumption.

The aggregated records of entries in the Journal activity (a record of all other activities, or programs, opened on the laptop) across all children with XO laptops in Paraguay provide quantitative data to corroborate my observations. Fully 32.6% of the program-opening events recorded in the XO laptops' Journals were "unrecognized," which included both Wine (a program that can load and run Windows programs) and Gnome (an alternate

Linux desktop environment), from which children launched games or video players. The next most popular program opened was the XO laptop's browser, which accounted for 13.4% of events recorded. Jukebox and Tam Tam Mini, the options in Sugar for playing music, accounted for 5.78% of the events recorded—although this underrepresents music playing, as I witnessed children also playing songs through Gnome and Wine. On the "productive" side, the word processor and office suites, both frequently used in the classroom, together accounted for another 9.38% of events, and the Record program, which allowed children to take and view pictures and videos, rounded out the top five. Overall, these five types of programs, out of the 152 total programs logged, accounted for over two-thirds of recorded program-opening events, as summarized in table 4.2. Three-quarters of the rest were opened less than 0.2% each, and half were opened less than 0.02% each (often just once), making up a very long, slim tail.

Many of the teachers supervising these children were aware of their leisure activities on the computer. Still, not many wanted to control what children did in their free time, and some felt that any laptop use was teaching the children useful technological skills. This view is echoed by the literature on new media literacy and connected learning, though it has been critiqued by other scholars.[13] "Outside of the classroom," one teacher at a small rural school explained, "kids will listen to music and play"—and that, in her view, was okay. However, their leisure use initially made her view

Table 4.2.

The five most popular types of programs opened, as recorded in Journal in February–August 2010.

Program name	Recorded openings (#)	Recorded openings (%)
Unrecognized (Gnome, Wine, etc.)	58,828	32.60%
Web browser	24,228	13.42%
Word processor, office	16,931	9.38%
Record (camera, webcam)	13,515	7.49%
Music (Jukebox, Tam Tam Mini)	10,429	5.78%

These include the browser, office suite, camera/webcam, music players, and unrecognized programs, which together accounted for 68.67% of events recorded.

the laptop as a little toy: "Last year I thought it was just a toy, not a tool for classes—more for games and music." A teacher at another small rural school was more disapproving when she described what children at different ages did and how this had changed over time. "The older children visit prohibited [pornographic] sites," she told me, grimacing. "The younger children are interested in games. At first, they were interested in downloading rude/gross [*grosero*] things, and they would show me. After that, it was music and video clips."

On the one hand, this again highlights the agency of those using XO laptops every day to choose the uses that they found most captivating—an agency that was lacking both in narratives from OLPC and those by scholars critical of the project.[14] Clearly, children were finding their own uses for the machine in spite of what OLPC had planned for them. On the other hand, we will see that XO use was instead channeled toward capitalist and consumptive ends by large media corporations—similar to televisions and radios, both ubiquitous in Paraguay. Similar use has been condemned among professionals and many middle-class parents in the United States as "screen time," in part because parents fear that the agency in choosing to consume this content lies not so much with their children as with the media corporations that design content and related merchandise to be as captivating as possible to impressionable youth.

1. Video Games and the Limits of Charisma

Video games in particular were wildly popular among the children who did use their laptops, boys and girls alike. Almost all children I interviewed and observed who used their laptops at all had their favorite games, whether these were an online Barbie dress-up game or a car-racing game played in the Windows program loader Wine. For instance, one seventh-grade student explained to me, "I play Flash internet games. My favorite is Barcelona [a soccer game]; it has matches and two penalties and so on. We use Wine a lot for games." Other favorites that I saw and heard about during my fieldwork included Vascolet, Super Vampire Ninja Zero, Mario Bros. (which came with Wine), Tux Kart (an open-source version of the popular Nintendo game Mario Kart featuring Tux, Linux's penguin mascot), and various soccer-playing games. Before my fieldwork started, the classic first-person shooter game Doom circulated into, and then out of, popularity as well; most adults who mentioned it roundly condemned its violence.

At first blush, this video-game playing appears to corroborate the vision of OLPC and constructionist learning. Both Negroponte and Papert discussed the benefits of video games—indeed, video games were central to the play that Negroponte, Papert, and others wanted to happen on the computer, as we saw in chapter 2. "There is no question that many electronic games teach kids strategies and demand planning skills that they will use later in life," Negroponte wrote in *Being Digital*, though without providing evidence for his claims.[15] Papert had similarly glowing but uncorroborated things to say about video games. "Any adult who thinks these games are easy need only sit down and try to master one," he claimed. "Most are hard, with complex information—as well as techniques—to be mastered."[16]

In this view, video-game playing would appear to at least partially fit the charismatic visions of the project—but one assumption that these OLPC leaders made was that video games would lead children into programming and other, more sophisticated computer use, rather than being rather limited ends in themselves. It is true that certain kinds of video games can at least temporarily improve reaction times and spatial reasoning, but their connection to other kinds of learning is much more contested.[17] Some games—like some books, movies, television shows, or other media—have complex storylines and grapple with heady issues. Some, such as Minecraft (which was in an alpha version during my 2010 fieldwork and was completely unknown among my participants at that time, but had been played by a few by the time I returned in 2013), have since been hailed as pathways into programming.[18] These were generally not the kinds of games I saw being played among children in Paraguay, however. The *fútbol*, car-racing, and side-scrolling games that most of them played were great fun, but, as one teacher said, "they do not require much mental processing. ... They are very easy, like watching television."

Those involved in the project in Paraguay Educa—in particular, the caretakers and teachers in charge of both overseeing children's laptop use and facilitating their learning—were divided on the value of video games. At one extreme, one school director optimistically explained, "For me, no game is a negative influence. If the children have the right orientation, if they are motivated by the games, that is positive. From the game, our task is to make learning meaningful. There are no bad games; it is just a matter of how you approach them." Others were unequivocally against video games, though generally not for the caricatured reasons that Negroponte and Papert depicted. "I personally do not like the little games they play instead of doing homework or talking to their parents," one teacher told me. "I particularly do not like the violent games." Many agreed that, up to a certain point, games helped children become more proficient in using their laptops but also felt that most children reached that level of proficiency fairly quickly and that further game playing was leisure, not learning. One teacher and mother of five children, three of whom were in the laptop program, quipped, "The educational games are like studying, and they do not like them"—and this included constructionist programs such as Scratch, eToys, and Turtle Art. Thus, even though video games were captivating to children, this use did not seem to fit OLPC's charismatic vision for them as a playful path into programming.

2. Music, Videos, and the Limits of XO Storage

Similar to video games in popularity were music and videos, which, in 2010, children generally downloaded to their hard drives throughout the school day and played offline on their XO laptops. During fieldwork observations, I found that music in particular was nearly ubiquitous. Some children played music in the background off their XO laptops while they surfed the internet during recess, while they walked home from school, while they played at home, and even while they worked on schoolwork in class (if their teacher was lax enough). In this way, they added a soundtrack of their choice to all of these activities—much as many young people in the United States have done via phones, MP3 players, and the Walkman before them.[19] One student said that she liked to download music to dance to at home. One mother commented, "They play, watch TV, or use the computer if we are at home to listen to music off their pen drive [USB memory stick]. ... They just like music and soccer."

Rap-like *reguetón* music was most popular, and children often mentioned Puerto Rican musician Daddy Yankee as a favorite. During my 2010 fieldwork, I also heard a lot of Shakira's "Waka Waka" theme song for the FIFA World Cup, as Paraguay was gripped with World Cup fever in celebration of its first-ever ascent to the quarterfinals that winter. I also heard many other pop songs in both Spanish and English, including familiar US-based pop stars such as Lady Gaga, Miley Cyrus, and Michael Jackson. Many of the videos that children played and shared were music videos for their favorite songs.

Though music was fairly ubiquitous all around town, not all parents and teachers liked that their children used their laptops to play it. A parent of a second-grade boy complained to me that her son was not interested in learning how to use the laptop; he only wanted to be able to connect his XO to the house internet connection to download songs at home.[20] "He just wants to listen to music or play games," she told me. "That is all that interests him." One teacher objected to the distraction a new song could make in the classroom: "The main negatives are the music and video websites. They will get distracted and not pay attention to class or their work. This is especially when a new song is available; then they will focus on it."

These large media files also quickly filled up the one-gigabyte storage capacity of the XO laptop, leading to the uninstallation of programs that teachers wanted to use in class—as well as the deletion of their own creations. Coming as I did from a culture of obsessive data backups, these children's cavalier attitude toward data loss was surprising to me. Not only were many children unperturbed when Sugar software updates deleted all of the school projects and other work on their laptops, some actively deleted them themselves, along with activities that came with the machine, to make space for memory-hogging media. However, a few children told me that they had lost the small amount of interest they had in creating on the XO when their projects were accidentally deleted in a software upgrade, suggesting that data impermanence could accentuate practices of consumption over creation. If children could not trust that their projects would not be accidentally deleted, why should they bother putting lots of time into them?

3. Pornography and the Limits of Self-Regulation

Finally, I heard stories about one of the most worried-about topics for children on computers: access to pornography. Though every school had an internet filter in place to block as much sexual content as it could, no filter is perfect, and one technician told me that the logs on the school servers confirmed that not all pornographic sites were effectively blocked. I did not ask children about this directly, and none ever talked to me about it or showed me anything, but I did ask parents and teachers in interviews whether they had any worries about the internet, and many discussed their concerns about "prohibited" or "inappropriate" content. No parents knew whether their own children were viewing anything in this category, but a few teachers and trainers had encountered pornographic materials on students' laptops. One trainer (employed by Paraguay Educa to encourage classroom laptop use) told me about catching a student showing cartoon pornography to friends not once but three times.

Pornographic content was ubiquitous enough on the internet that at times it was hard to avoid, even with a filter. A fifth-grade teacher described an incident that had happened during a lesson in class, when she had instructed students to research parts of a flower on the internet. One student found a picture of a penis when he searched for "pistil" (possibly misspelled) and gleefully showed it to the students around him—and soon the whole class was clamoring and yelling to see. The teacher tried to make students put away their laptops and have an impromptu discussion of the similarities between flower parts and body parts but could not regain control over her classroom. In another incident, a teacher tried her search ahead of time but for something far less suggestive. She explained, "I was searching for information on the sense of smell, and suddenly these rude things appear! If a child saw that and asked, 'What is that?' their innocence would be lost very fast. So that is why I want to monitor; that is my fear. I know you have to learn these things in life, but all in due time. I do not want them to hurry."

This teacher's thoughtful effort to balance children's need to learn the facts of life in due time with a desire to maintain their "innocence" as long as possible is in stark contrast to the glib cyberlibertarian remarks that several OLPC employees and contributors had made on the topic. Seymour Papert claimed in a 2006 interview that children would simply moderate themselves in looking at salacious content (which the interviewer called "weirdness") on the XO laptop: "We envision 100 million laptops being in the hands of children in a few years' time. It is impossible for us to even think about moderating what all these children are doing. ... The proper kind of moderator is the children themselves. The children themselves should be the control over the best use of the computers, and preventing what you call weirdness."[21] Similarly, in response to a scandal involving pornography watching in a Nigerian OLPC pilot program in July 2007, the mostly male OLPC community collectively shrugged and said that pornography is a large part of the internet—and because children will encounter it eventually anyway, why try to regulate it?[22]

These opposing viewpoints characterize two sides of the debate on pornography and children, as well as on the regulation of children's media more broadly. On the one hand, parenting organizations such as the American Academy of Pediatrics match the view of the parents and teachers quoted here in their focus on protecting children from what is considered harmful media content, however that is culturally defined.[23] Methods for implementing this protection are sometimes individual but often include government regulation or technical solutions such as content filtering. On the other hand, Papert and OLPC's view matches the stance often taken in the US technology industry and championed by the digital-rights group Electronic Frontier Foundation that any attempt at censorship—even of content that is widely considered reprehensible or on behalf of audiences that are widely considered vulnerable—is technically bound to fail due to the abundance of technological workarounds (as the failures of Paraguay Educa's filters show) and legally could lead to a slippery slope that begets more censorship.[24]

Some who hold this latter view advocate for self-regulation, as Papert did in the interview above. Elsewhere in his writings, he indicated a mistrust for most parents and other adults to understand children's interactions with technology.[25] Others are more equivocal on parents' roles, suggesting that parents could monitor and restrict their children's media exposure manually, in accordance with their personal values. Both of these approaches are neoliberal in that they are decoupled from the state, either trusting the market to follow popular sentiment or placing the onus on the individual to self-regulate. The ongoing debate on these matters shows that there are no perfect solutions.

Media, the Internet, and Agency

Although this media consumption did not match the charismatic promises that OLPC had spelled out for the project, there were certainly elements of learning taking place in how these children retrofitted a machine not designed for media-centric use. In keeping with what OLPC imagined that children would want to do with computers—learning to program, creating content, exploring Wikipedia and other information online, or connecting with one another—the XO was loaded with educational software and was not designed for playing video or audio. The version of the software in use during my fieldwork in 2010 intentionally could not run content in Flash (a fairly common format in 2010 for online videos, music, games, and interactive websites), did not come with a preinstalled video player, had low-quality speakers, had a one-gigabyte hard drive that was wholly inadequate for song and video file sizes, and by default played audio through either the Jukebox or the Tam Tam Mini music suite programs, both of which were designed to engage children more in creating their own music than in consuming it. Moreover, in Paraguay, YouTube was added to the list of blocked sites relatively early on (before my arrival), removing one of the largest sources of video on the internet, because children were using it to watch videos that parents and teachers found much too violent.[26]

Thus, children interested in consuming media on their XO laptops had to install additional programs to enable this use. Their source of information on this was generally the same as their source for media: the internet. Indeed, for all of the frustrations that children (and teachers, as we saw in the previous chapter) expressed about using their laptops, some were excited to have access to the internet. Most only had access to an internet connection at school, which teachers and directors said had a nice side effect: it increased attendance. At times I even found children sitting just outside school buildings after school hours or on weekends, downloading files. One teacher told me, "Students who did not come to school regularly now do, so they can get on the internet and download games and music." Even so, this teacher was not sure about the long-term effects of this media exposure. "Games have their place," she explained, "but it should be a small part of their lives, so they can practice mathematics, which just is not as fun." During my 2010 fieldwork, internet connections proliferated across Caacupé—first in the twenty-six Phase II schools, then in the public plaza downtown, to the eager anticipation of Phase II students. I asked them why they were so excited, and they unabashedly said it was because they could download and play games and music. A few schools who turned their access points off during school vacations were pressured by children to leave them on.

How did information travel from the internet to children across the town? The specific sources that children found most useful in learning to retrofit their machines included a number of XO-dedicated websites that explained how to install workarounds and provided installation files for download. Many of these sites were hosted in Uruguay, where a large, active community of software developers provided the countrywide project with volunteer technical support (e.g., http://rapceibal.info), training classes, and lots of how-to websites in Spanish. Two of the most popular websites during my fieldwork, which I frequently saw in my observations, were Portal XO (http://www.portalxo.org) and XO Planet (http://xoplanet.blogspot .com). These sites were supplemented by (and sometimes copied) US-based websites created by computer enthusiasts who bought first-generation XO laptops during OLPC's Give One, Get One programs during Christmas 2007 and 2008 and posted their own workarounds for the XO laptop's hardware and software limitations. Several of the games popular in Paraguay, including *Doom* and *Super Mario Bros.* (via Wine), were originally ported to the XO by US-based programmers.[27] Others were developed closer to home, including several versions of *Vascolet* (which was sponsored by Nestlé and starred a character the company had developed in 1974 to advertise Nestlé's chocolate-milk powder across Latin America) and *Super Vampire Ninja Zero*, both created by Montevideo-based Batovi Game Studios (http://www .batovi.com). Some of the oldest and most technically savvy children discovered these sites through web searches and then showed other children the tutorials. They would learn, and then teach one another, how to install alternate desktop environments, such as Wine or Gnome, to play video games and open-source media players to watch videos and listen to music over the XO's small speakers.

Another useful source of information was Paraguay Educa's trainers and technical staff, all of whom became very familiar with the XO laptop as part of their jobs. Children quickly learned that staff members were useful sources for tips on what to do with an overly full Journal or how to download and install new programs. One seventh-grade student had learned how to use the "rm" (remove) Linux shell command from a technician as a way to clear his computer's memory from the command line. In an interview he told me, "I just use Terminal to delete my Journal and install Flash Player so videos will go faster. In Gnome, I use Virtual DJ for music." Two sixth-grade boys whom I interviewed together attributed their knowledge of where to find videos to their school's teacher trainer and from visits by Paraguay Educa's programming team. They also told me that they wished the trainers would "unlock" YouTube.

It was from these sources that video games, movies, music, song lyrics, jokes, and more circulated from student to student, supplemented by

USB flash memory drives, locally called "pen drives," which a few wealthier children owned.[28] These drives allowed content to continue to spread even away from the school-based wireless network. One teacher trainer corroborated my own observations of how content circulated around the schools. "There are two or three students in every room that are really good at searching the internet and finding games and other pages you would never imagine existed," the trainer explained. "They share with the other students. They also know how to get around the school's firewalls and download music [some of them doing so with home internet connections]." In a group conversation, several other teacher trainers commented that new songs or games could spread throughout the schools where they worked within a day, starting before school, continuing at recess, and jumping to the afternoon session via children from the morning session staying late to surf the internet overlapping with children in the afternoon session coming early to do the same. Children, too, discussed the social sources of their media with me. "My classmates showed me *Vascolet*. They gave it to me on a pen drive," said one fourth-grade student in a larger rural school. A second-grade student in a large private school told me, "My favorite thing to do is download music in Wine. I do not know how to [do this] myself, but friends help me at school." Information also flowed between siblings and cousins, again generally from older to younger, which allowed it to jump between school sessions and schools as well.

In these ways, children could at least partially work around the limitations of the XO laptop to make it serve their media-focused interests. One trainer quipped that the same children who were good at downloading games and other content would struggle with using educational programs such as Scratch, one of the featured constructionist programs on the XO laptop.[29] "They are super smart only when it is convenient for them, when they want to—only when they are motivated," the trainer observed, with equal parts admiration and exasperation. "It is amazing. Kids who cannot read or write know how to download games, sometimes even in English. They memorize what to write, even if they don't understand anything—click here, copy and paste there, follow the links. They follow directions well. But if you give them directions on the computer for something *you* want them to do, they cannot do it. It is amazing."

His exasperation notwithstanding, this teacher trainer's comment highlights that in using their machines for media, these children were still engaging with the machine. Though it might not have involved using some of the more constructionist programs such as Scratch, Turtle Art, or eToys, they were nonetheless increasing their technological literacy. As one teacher put it, "For me it was not a failure; for me it was a success. ... Even if the child just learned to download only music or play games, that's something learned." As they sought workarounds to the limitations of the XO laptop and the filters on school internet connections, these children did use their XO laptops—but on their own terms. This highlights the agency of these children in choosing the ways in which they wanted to engage with the machine, even if their choices worked against the design of the XO laptop itself and ultimately did not uphold OLPC's charismatic promises of inspiring technically precocious children across the Global South to learn to program computers.

The Limits of Nostalgic Design

We have seen an apparent disconnect between the prototypical child that OLPC imagined and the actual children who received laptops in Paraguay. The source of this disconnect was nostalgic design. As chapter 2 explored, one of the reasons that OLPC's laptops were so charismatic for many in the world of computer and software development was because they resonated with the rosily mythologized stories that many told about their own childhoods, which tended to evoke the prototypical yearner child of Papert's writings. They seemed to imagine that these yearners existed in great numbers elsewhere—even as many described themselves as exceptions among their peers in their own childhoods.

This chapter shows, however, that XO laptops were not being used like the programming machines discussed in Papert's books *Mindstorms* or *The Children's Machine*. They were also not being used in the same way as preinternet computers and gaming/programming devices, such as the Atari or Commodore systems, which some OLPC contributors fondly discussed using in their own youths. Instead, the children in Paraguay whom I encountered in my fieldwork were using their XO laptops as many around the world use computers today. About half of them were just not that into their machines. They found using the XO unrewarding, whether because of the limitations of the laptop itself or because of the fullness of their lives without it. Another one-third engaged with their laptops as media-rich, internet-connected sources of entertainment. Despite the charisma that a specialized laptop with lots of unique features and educational programs held for its creators and others in the technology world, these children were using their XO laptops as they would use any other computer. In fact, they had to work hard to overcome the limitations of the laptop's design to enable this use—limitations put in place to make it cheaper, yes, but justified by referencing the learning that took place on the developers' much simpler childhood computers in a less saturated, less ubiquitous media environment.

This disconnect between OLPC's vision and Paraguay's reality suggests a larger shift in the way that computers are imagined and valued collectively, and the meanings we culturally attach to them. OLPC based its design on the kinds of programming-centric machines that were popularized in 1980s media and that first glorified hacker culture—a time when, not coincidentally, many of OLPC's contributors came of age or defined their careers. This command-line world was populated by a relatively small group of early adopters who could afford the expense of such devices and understood them enough to want to overcome fairly steep barriers to use. However, today's primary use of computers for leisure—whether in Paraguay or the United States, whether among children or adults—is centered on media consumption. Graphical user interfaces have long since removed many of the barriers to use that command lines had reinforced, computers are much more widespread, and content is much more media rich. In this environment, with these expectations, it is no wonder that children in Paraguay had little interest in exploring the components of the machine that resulted from OLPC developers' nostalgic design; such components struggle to hold the attention of children everywhere in the face of an internet-connected, media-rich world. For these children, nostalgic technology design held no charisma; it was not their rose-tinted childhood memories being referenced.

Media Imperialism and the Limits of Agency

Some scholars have critiqued OLPC's ideological imperialism, with a focus on analyzing OLPC's promotional materials or news stories about the project. For instance, Rayvon Fouché analyzes OLPC's "technology as salvation" rhetoric for its racist undertones, finding that the project employed a common trope of the white savior from the Global North who decides what is best for "lazy" brown people. Fouché posits that this trope could fit comfortably in the contemporary landscape of racial resegregation and the revival of beliefs in innate racial difference in the United States, but understandably chafed many in the Global South. Its individualized rhetoric, moreover, obscured the institutional racism that helped create the very social, economic, and technological marginalization that OLPC sought to "fix."[30]

Despite the fact that such analyses are insightful—especially as critiques of the culture of American technology development—they are also incomplete in that they do not include the other half of the story: the meanings that users constructed about these laptops. This chapter—which shows the extent of nonuse, breakage, and media-centric use of OLPC's XOs—brings back the agency of children that was lacking in narratives from OLPC, as well as those by scholars critical of the project. Both OLPC's narratives and early critiques of the project predicted that local cultures would be co-opted or erased by a brave new technological world created in OLPC's image.[31] At the same time, there were additional influences in children's lives that shaped their agency to use their laptops as media machines: the large corporations that sponsored content on the XO. They created the games, music, and videos that appealed to these children and in some cases, such as Nestlé's Vascolet, advertised to children through these games.

The potential influence of media corporations is largely absent from presentations and publications about constructionism and OLPC, which (perhaps purposefully) seemed to exist in a cultural vacuum where media influences had little role in children's lives and where learning through creative play was the natural focus. In this world, there was little to do on computers besides tinker with them and explore their inner workings; even the discussions of video games in Negroponte's and Papert's writings were divorced from any sense of branding or other corporate presence. This is laudable as a goal, but this chapter has shown that media corporations certainly had a presence in these children's lives through the XO laptop. In Paraguay, as in much of the world, the products of multinational companies are heavily advertised and well known, from *The Simpsons* to Michael Jackson to Coca-Cola. In the face of these companies' vested interests in steering children's leisure time toward particular forms of branded consumption that benefit their corporate bottom lines, it seems naïve (to say the least) to expect children to manage or critically assess, much less resist, these media influences on their own.

There's an interesting twist to this story. The OLPC leadership that I have focused on so far—the leadership responsible for popularizing the idea, designing the laptop, and setting up projects—was relatively silent on the topic of media. However, there was another branch of One Laptop per Child: the OLPC Association, which was built up around 2009 and 2010 in Miami, Florida, to help manage existing projects and ended up inheriting OLPC's trademarks when the original Boston-based group (which had become the OLPC Foundation when the association was created) dissolved in 2014. This branch seemed to not fully recognize—or to intentionally ignore—the disparity between the subversive, programming-centric laptop use that OLPC initially championed and the more lightweight, media-focused use that I witnessed and that corporations would likely encourage. Moreover, the association had the benefit of considerable goodwill built up behind the OLPC brand—a point that worried many who had been part of the original OLPC community.

In July 2011, the OLPC Association announced an official partnership with Nickelodeon in a contest "to design multimedia about improving the environment."[32] A Paraguayan girl won this contest and was awarded a trip to the Nickelodeon HALO Awards ceremony in Los Angeles in September 2011.[33] In 2013, during my follow-up fieldwork, she showed me her submission: a montage of Nickelodeon characters that she put together in Scratch to illustrate a typical day in Paraguay—a brand-oriented submission that Nickelodeon apparently rewarded. Though few could argue with the goal of improving the environment, Nickelodeon heavily branded the contest and undoubtedly enjoyed public-relations benefits across Latin America for little cost or effort as a result of the competition.

In this way, transnational corporations could move into OLPC projects across Latin America and take advantage of this new market of young protoconsumers. Though most children in Paraguay already had access to a television and were surrounded by the music and products of media corporations before the laptop program started, XO laptops allowed unsupervised media consumption by more children, at younger ages, via avenues that their teachers and parents often sanctioned as educational. Thus, the agency that these children exhibited in choosing to download Vascolet video games or Daddy Yankee songs is tempered by the media imperialism of the multinational corporations that own these brands and often create this content with US consumers in mind.[34] This content was not new with the XO laptops but found another avenue to reach children through them—an avenue billed as educational.

Claims of media imperialism are complicated. On the one hand, critical communication and media scholars have shown that American movies, stars, and media models (such as reality television or coordinated media properties) seem to override local media practices. These models are problematic in that they often presume a one-way flow of media products from a culturally aggressive United States to passive and often Othered recipients elsewhere, rather than acknowledging that such uptake is an active cultural bricolage of meaning making, aspiration, and agency—and, moreover, that this bricolage has been ongoing between various cultural groups for hundreds, if not thousands, of years.[35]

Still, an element of media critique creeps back in when we consider children's media in particular and its contentious role in families and communities around the world, the United States included. The cultural battles over pornography indicate that many feel that the stakes are high and that there are no clear answers for how best to moderate children's media access. Moreover, although these media properties may be tailored for US audiences, with US cultural values in mind, it is ultimately media corporations that profit from their dissemination. If we shift the frame from one of media imperialism to consumerist critique, it makes the stakes of advertising to children via "educational" devices clearer: these companies are profiting from their access to these devices.

Charisma and Social Change

Papert's writings on constructionism and Negroponte's talks on the precursors to One Laptop per Child tended to focus on the few children who enthusiastically engaged with the systems they were prototyping in classrooms or homes; they often entirely omitted the rest of the children in the classroom from their narratives. The chapter brings balance to this picture by avoiding the temptation to focus exclusively on the engaged children, those rare emblems of success who make it easy to confirm one's theories. Whereas we will examine such children (and their social contexts) in the next chapter, here we examined the practices of the vast majority of children in the program.

We may question the expense of a program that did not engage half of its intended beneficiaries, exhibited breakage rates that were much higher than expected, and required considerable retrofitting to enable desired uses. But the consequences of charisma's limits in Paraguay go beyond cost to question who is seen as responsible for creating "development." Because a charismatic technology is supposed to make social change easy, even inevitable, then it follows that the key for OLPC—and many other charismatic projects in education and development—to create the conditions necessary for change is mere access to its technology. Wooed by some of the same charismatic promises, much of the early literature on the digital divide similarly focused on questions of access.[36] For OLPC, the idea was that if children just had XO laptops in hand and if they just bothered to engage, the machine's charismatic potential would unfurl and carry the children along with it. But what if a child failed to engage—or failed to engage in the right way? Unspoken in this charismatic story was an individual responsibility to use the technology provided in ways that fit expectations, which would not only benefit the individual but also bring about collective cultural uplift. In this model, the Paraguayan children I followed would have nobody to blame but themselves for not buying into the charisma of the machine in order to achieve these miraculous results.

This chapter has challenged this characteristic of charismatic technologies by showing how charisma was limited from many angles among its primary beneficiaries in Paraguay: limited technically, limited socially, and limited historically. Instead, the agency of the children themselves came to the fore: some children rejected the machine, while others retrofitted it for the uses that they found more captivating than the ones that OLPC had wanted to foster. At the same time, this chapter also highlights the centrality of the products of media corporations in these children's lives and questions the influence these corporations are allowed to have, including on the XO laptop.

5   The Learning Machine and Charisma's Cruel Optimism

For the last two hundred years, we've assumed that the way you learn is that you go to school and that somebody teaches you. And it's the first time that learning can happen without teaching. That sort of cusp happened because of computing, because of the internet, because of interactivity, because of a lot of things. So, you add that to the fact that the world right now is in a situation where as many as three [or] four hundred million children around the world aren't getting any education; they don't go to school. And so, you have to leverage the children. You can't just train more teachers and build schools. You've got to go back and say to yourself, "Can the children be part of this rather than just the objects of teaching?"

—Nicholas Negroponte, quoted in Kleiman, *Web*

With breakage, disinterest, and media consumption dominating the laptop use I saw among children in Paraguay, it may seem that the charisma that so motivated the technology world to invest in and support One Laptop per Child—and the social imaginaries that undergirded that charisma—simply did not translate to the Paraguayan children who received OLPC's XO laptops. However, juxtaposed against OLPC leaders' claim that the laptop could be a universal learning tool were also statements that suggested that the project could still succeed even if it did not reach all children, as long as it reached enough.[1] Even if only a small number of children were inspired, this small number might go on to not only transform their own lives but, through their entrepreneurialism, transform the economies of their regions or even whole countries.

"We need to reach the most children possible and leverage them as the agents of change," Nicholas Negroponte said in a publicly archived 2008 email to OLPC contributors.[2] He echoed Seymour Papert's idea that constructionist learning through computers would give agency back to children to determine their own educational paths. "Increasingly, the computers of the very near future will be the private property of individuals, and this will gradually return to the individual the power to determine patterns of education," Papert explained in *Mindstorms*.[3] In a 2006 interview about OLPC, Papert reiterated a belief in individualized learning: "There are many millions, tens of millions of people in the world who bought computers and learned how to use them without anybody teaching them."[4]

This implicit individualism was one avenue by which OLPC would overhaul education; it took parents, teachers, and broader cultural systems more or less out of the equation. After all, even with ready access to computers, relatively few children in the Global North identify with the social imaginary of the technically precocious boy and become the rebellious tinkerers that populated the OLPC contributor community and the hacker world from which it grew—but this small number appears to be enough to help drive the engines of innovation in Silicon Valley.

This individualism also cropped up in Papert's parallels between learning mathematics and learning a spoken language. When Papert discussed language learning, the only actors present in the story were the learner and the words they were absorbing.[5] Absent was any indication that a strong motivator for learning a language is the wish to communicate needs and desires with other people—absent, in fact, are the other people speaking these words. Similarly absent was any recognition that as miraculous as children's language learning can seem, it is nonetheless scaffolded by adults, who offer definitions, corrections, and models that help structure children's rapidly expanding vocabularies. Instead, Papert claimed that a child plunked down in France for a year "spontaneously" learns to speak French, rather than being actively (if informally) taught it.[6] Even when the social inevitably crept into Papert's narrative—particularly when he discussed "culture"—the language he used to describe it was abstract, even epidemiological. His metaphors of "germs" and "seeds" of mathematical learning within a "landscape" of ideas reduced the social world to a toolbox where children might encounter gadgets that help them learn mathematics and logic, stripping away the complex social motivations and interactions that constitute culture.[7]

One challenge of this vision is that it situates the locus of both learning and social change on the individual. It will be individual children who identify with the charisma of the XO laptop, and through the force of their inspiration, they will uplift their communities to the freedom and prosperity that OLPC promised. Even though there is something inherently unjust in accepting that only a few children may benefit from a charismatic technology—which children, under what circumstances, and with what effects on equity?—there is another element that is even more problematic. In particular, the project's focus on student-led learning, drawing as it does on imaginaries such as the technically precocious boy, does not account for the critical role that various social worlds and institutions—including peers, families, schools, and communities—play in shaping a child's educational motivations and technological practices.[8]

In this chapter, we will explore in depth the lives of several Paraguayan youth (older children and teenagers) who would generally be considered OLPC success stories—those few exemplary children for whom the charisma of the laptop did seem to resonate—and what became of their inspiration. Doing a thorough and systematic search to identify these children was a focus throughout my fieldwork. I watched for them during my observations in the ten Phase I schools with laptops in 2010, and I asked students, teachers, teacher trainers, and technical staff which youth seemed especially proficient or inspired. My field notes and their overlapping answers triangulated on a small set of youth that were indeed engaging with their laptops in creative ways that contrasted the disuse or media-centric use of most of their peers. Casting a wide net, I encountered about forty youth in 2010 who were engaged in these more creative uses: they helped their teachers use laptops in the classroom, photoblogged, learned basic repairs from Paraguay Educa's technical staff, and explored the basics of programming in one of the three constructionist block-based programming environments (Scratch, eToys, and Turtle Art) on the XO laptop. Almost all were in the sixth and seventh grades in 2010 (aged twelve to thirteen), the oldest two grades in the program that year. If I restricted the practices to only count programming—the kinds of uses that OLPC was most interested in fostering—I found around eight youth engaged in this in 2010, all of them using block-based programming environments. In 2013, I again followed up with these youth, most of whom were no longer engaged in the computer in this way, and I also interviewed and observed another eight high-school students who had started programming through a Saturday code club sponsored by Paraguay Educa.

Although the uses to which these youth put their laptops were not (or at least not yet) the sorts of activities that technologists would generally identify as sophisticated—in both 2010 and 2013, they were putting together basic code with great difficulty—those uses were still distinct from the consumptive practices of their peers and in line with what constructionism and OLPC envisioned. In the constructionist frame, children's motivation was a matter of the child finding something irresistible about engaging deeply and meaningfully with the laptop, which in turn came from the natural creativity and technical inclinations that all children are born with. Indeed, if one spoke to these youth casually with this frame in mind, they might well appear to fit this assumption. But in exploring their lives in more depth— watching how they used their laptops, observing their home environments and daily routines, and interviewing their parents or caretakers—I uncovered a much more complicated story.

In this chapter, I consider the ways in which this learning was situated within and motivated by the social worlds of these youths.[9] In the process, I tease out the power, and the limits, of these environments in creating the kinds of futures that OLPC and Paraguay Educa wanted to create for them. We will see that even when the project was ostensibly successful—that is, when its charismatic promises appeared to be fulfilled—its individualist focus failed to shift the larger structural inequalities that worked to marginalize these youth. This led to a retrenchment of gendered, socioeconomic, and linguistic inequalities rather than the sweeping cultural uplift that the charisma of the XO laptop had promised.

Manuelo and Elisa, the *Scratchero* and *Scratchera* Siblings

By far the most celebrated of these youth during my 2010 fieldwork was a seventh-grade boy, whom I will refer to as Manuelo. Though reserved around many adults, including his teacher and me, Manuelo was very social with his siblings and was especially close to his sister Elisa, who was one year behind him. Almost the same height and with similar features, right down to their shy smiles, the brother and sister were often mistaken for twins. In 2010, both were particularly enthusiastic about Scratch, a block-based visual programming environment in which "Scratchers" can "snap" together code blocks to define the characteristics of background images ("scenes") and characters ("sprites," borrowing language from game programming). They can also back up their projects and see the projects of other Scratchers via the Scratch website.[10]

The siblings worked so closely together on joint projects in Scratch that they could not tell me who contributed which parts. One project that they each showed me in interviews was a game they had created together. In this game, a player had to navigate a small dog character through a set of increasingly difficult vertical mazes—sliding left and right and jumping up to higher platforms from the starting point at the bottom to the exit at the top—while avoiding arrows that shot diagonally through the scene. Though lacking many elements of games such as scoring, start-up scenes, end scenes, cut scenes, save points, or story, their game was quite complex. It had nine levels, each consisting of a one-screen maze, which got progressively more filled with arrows. Elisa boasted that nobody had been able to clear even the third level. They not only designed the dog character and the background scenes but programmed the game logic as well, using Scratch actions such as "walk," "jump," and "turn around" to control the dog's movements on the platforms and "disappear" if the dog collided with an arrow. These characteristics mimicked aspects of side-scrolling video games such as Super Vampire Ninja Zero and Nestlé's Vascolet, which were popular in Paraguay at the time.

Both siblings told me that they had created this game together out of whole cloth, rather than building off (or "remixing" in Scratch terms) a sample project included with the program or a project created by another Scratcher from the Scratch community website. Elisa had an account on the site, but she said she only created it to back up their work after a software upgrade had deleted some of their earlier projects. Both told me that they had learned how to use Scratch code blocks largely by opening other projects and seeing how they worked. Manuelo also described learning about the code blocks by "just trying them all one by one."

Manuelo told me that he and Elisa had decided to make games "because that is what people like the most—it is what they use." Their mother, María, had a somewhat different account of why they decided to make their own game. María was a teacher at the same large public school her children attended. A smart, dynamic, and articulate woman in her late thirties, she was always ready with a smile or quip and took her job as an educator very seriously. She had become the most enthusiastic teacher in the whole project and also a staunch advocate for her talented children. As a teacher, she had learned about the various programs available on the XO laptop in teacher-training sessions and then from the trainer in her school, and she later introduced Scratch and other programs to her children. With her encouragement, they soon outstripped her knowledge, and she sometimes purposefully played up her ignorance to get them to explore more. "To find out what they were doing," she explained in an interview, "I used to pretend to be a fool and ask them, 'How do you use this?' And they said, 'Oh Mom, come, and I'll explain,' and then I would say, 'Oh, that's how it is—and here's another way!' ... Then I would say, 'Oh, how nice, how good, what else?'"

In our interviews, María explained how she had modeled critical thinking and a good work ethic from an early age for her children and tried to do the same for her students. "From when they were young and started to think, I planned for them to be near me and inculcated to work like I work," she said. "My kids love Scratch. [It's because] from when they were really young, I would make sure they had something educational to play with: books, periodicals, cut and paste art from young ages. I was working in a school, so I was interested in what was educational." At the same time, she did note many children's widespread interest in music, videos, and video games. "I always asked myself this question: why do so few children use the XO pedagogically? And why do many children use it just to watch videos and listen to music or play games?" She recognized that her own children were in this group. "At first," she said, "when I didn't know what [my children] were doing [with their laptops], they'd download music and videos. But we blocked some of those in school, and we don't have internet at home," which cut down on media consumption. She then explained how she encouraged her own children to approach games and other media in a critical, reflective way, thus developing skills that she felt would be useful throughout their lives: "The children love the games—even more because of the cartoons they see on television. They're idols, and they internalize them. But I always have them reflect, the same as with videos. I tell them that videos and music will go out of fashion, but what I am going to teach them will keep."

María was not against video games entirely. In fact, on the family's desktop computer at home, which had been purchased four years earlier when María's eldest daughter entered college, there were several emulated

Nintendo games that her children loved to play. But she especially disliked violent video games and limited her children's use of with them. "There was the wave of popularity with Doom, which was a very aggressive game," she told me. "I told my children and students that it was too aggressive and had them reflect on it." And she felt that even nonviolent games should be used in moderation; she specifically encouraged using the XO laptop for reading, drawing, and other forms of creating.

Even with his mother's influence, though, Manuelo did use his XO as a media machine as well as for programming in Scratch. I heard all about it from his more gregarious fourth-grade brother, who told me, "I learned about Daddy Yankee [the popular Puerto Rican *reguetón* musician] from Manuelo. I also have a cousin at [a private school nearby] who tells me everything." He went on to describe how his older brother taught him to install video games and was his source for where to find movies and other media for the laptop. "I play with friends or my sister or brother—he gave me this game," he told me, pointing to the Nestlé game Vascolet—a side-scrolling game featuring a character who drinks Nestlé chocolate milk to gain strength—on his laptop.[11] "I also like to download music; I learned about these sites from my brother." He showed me a hard drive full of songs and cartoons, such as *The Simpsons*, and explained that he often had to delete other programs to make room for more.

Similarly, Elisa unabashedly described how she especially liked listening to music on her laptop. She said she had learned about the content she could find online from television. "I knew about the internet from TV—about websites," she explained. "From the TV I learned about YouTube and music, but I haven't actually connected with someone from another country. I have email, but I don't use it." In our interview, she went on to demonstrate how she could search for the lyrics to a popular Lady Gaga song, copy them, and translate them into Spanish on http://musica.com . "Here's a list of songs, and this is the video page," she demonstrated, searching for "Lady Gaga" on the site. "But YouTube is blocked at school, so we download video instead." Elisa, Manuelo, and their younger brother were engaged in what education scholars call "connected learning"—learning that was grounded in their interest in the video games and music available on the laptop, influenced by the social world of their peers.[12] Their use transcended the consumptive practices of their peers with the help of their mother, María, who provided scaffolding to steer their laptop use toward more educational ends.

María, her husband (who in 2010 worked long hours as a truck driver and in 2013 worked for the city), and their five children lived together in a house outside Caacupé. Despite interviewing and observing much of their family many times in 2010 and 2013, I never visited their house. María told me she was too embarrassed by her house to have visitors, because it was, as she said, "only half built." Instead, she showed me pictures of the exposed rebar where cinderblock walls would eventually be erected and the painted cinderblock rooms with her children's bunk beds (off which one of the family's XO laptops met its demise). But despite their humble living quarters (at least in María's opinion), the family was one of the few in the area that had invested in a desktop computer, which they purchased when the family's eldest daughter began college in 2006. This was rare: in a 2010 survey of all Caacupé teachers, I found that even among this relatively well-educated and often better-off group, only 25% owned a computer other than the XO laptop; estimates of computer ownership across Paraguay in 2009, the year that Paraguay Educa distributed the first batch of XO laptops, were closer to 10%.[13] As a result, Manuelo, Elisa, and the rest of the family were more familiar than many with the capabilities and limitations of computers. Moreover, María had a very clear vision for the role that technologies should have in her children's learning and play and felt that computer use was something that the family would discuss and even participate in together. As she explained in a 2013 interview, "We do not have much comfort, we do not have much material wealth, but we do have spiritual and family [wealth], and we are very united. My children they are very humble; they don't have branded clothes. ... We explain that in time, when they are professionals, they will have what they want, but for the moment, they will have what they can. And they listen well. I tell them that beauty is not what you wear, but what is inside."

Manuelo and Elisa both told me that they did not help their teachers much with their XO laptops because their teachers were not very interested in reconfiguring their classrooms and curricula to use them. Instead, the siblings came to the attention of Paraguay Educa through their mother, who had become an ally for the project. Paraguay Educa was duly impressed with them, especially Manuelo. They invited him to attend a Saturday hacker class taught over a couple of months in spring 2010 by a former OLPC programmer who had come to Paraguay to help with the project.

When I asked María why Elisa did not attend as well, given how closely the two siblings worked together, María said that she did not know if her daughter was invited and would not want her daughter to be the only girl in a classroom of boys anyway.

There were other instances of differential treatment: a Paraguay Educa staff member presented Manuelo and Elisa's game at the annual Scratch@ MIT conference in August 2010 but only included excerpts of an interview with Manuelo explaining it.[14] Two years later, in June 2012, Paraguay Educa staff arranged for Manuelo to travel to MIT himself to attend workshops for youth at the Scratch@MIT conference. Still, there were also some opportunities extended to both. Paraguay Educa arranged for Manuelo, Elisa, María, and twenty-two other Caacupé residents to receive four years of English lessons—lessons they were still taking during my follow-up fieldwork in 2013. In 2011, Paraguay Educa recruited the siblings and a handful of others to be *alumnos dinamizadores* (student facilitators) and encouraged them to teach their peers, even though neither sibling expressed much interest in doing so when I asked them about it. In 2012, after it had run low on funding and had laid off most of its teacher trainers, the NGO started an XO Evolution club that met on Saturdays and invited the siblings to that.

I considered Elisa just as precocious as her brother, and many of their projects were joint efforts. In fact, I thought that Elisa showed more initiative and creativity in finding novel and interesting resources on her laptop: whereas Manuelo focused most of his creative energy on the siblings' joint projects in Scratch, Elisa found a website that gave basic Japanese-language tutorials based on her interest in Japanese anime and manga, which she supplemented with the occasional practice she got with her school's guest mathematics teacher from Japan. She also liked to translate English lyrics to Spanish to find out what her favorite songs said. She loved the attention she got from Paraguay Educa as a *Scratchera*: "It makes me feel valued," she explained to me in 2010. However, a staff member at Paraguay Educa called Elisa a "conventional" thinker, while Manuelo "showed signs of genius"—a designation he seemed to receive because he resembled the social imaginary of a technically precocious boy that Paraguay Educa recognized, even if the siblings and their family did not. This imaginary was familiar to the internationally educated elite in Asunción who ran Paraguay Educa (and, indeed, was embodied by several of its technical staff), though I did not find evidence that it was broadly legible among students and teachers in Caacupé, where different expectations and imaginaries operated. As a result, even though much of Manuelo and Elisa's work was jointly created, Paraguay Educa extended more, and more exciting, opportunities to Manuelo than to Elisa.

When I returned for follow-up fieldwork 2013, Manuelo's and Elisa's computer uses had changed considerably. In those three years, smartphones and laptops (other than XO laptops) went from very rare to quite common in Caacupé and across Paraguay. Elisa had her own, an Acer, which she preferred to her XO because, as she said, "the XO is more for kids" and had not kept up with technological changes. "It's like fashion," she explained. "You cannot always dress the same way. ... The XO does innovate but little by little, and children lose interest." She used the Acer for reading, writing, playing some games, and accessing the internet—but she had lost interest in Scratch and had not even installed it. Elisa chalked up her 2010 interest in Scratch to novelty: "When you have something new, you spend days and months, even years, finding out if it interests you. But if it doesn't in the end, you just leave. I was interested [in Scratch] because I thought one day we would have to learn to program, and I didn't want to be the idiot who knew nothing. But now ... sometimes I go teach the younger kids [at Paraguay Educa's XO Evolution club], but otherwise I don't use it, and most Saturdays [when the XO Evolution club meets] I'm busy with friends, catechism, or other things."

When I met Elisa for her interview in 2013, she was leading a team of six peers in painting a mural on the wall facing the school courtyard. She had designed it herself, based on a drawing she had found on the user-generated art website DeviantArt: a daydreaming boy, eyes huge like a Japanese anime character. When I asked Elisa what she wanted to be when she grew up, she said she wanted to be a civil engineer working at the Itaipú Dam. Manuelo was less sure, but when he started high school, he decided to study the same thing—which, according to María, made Elisa "furious, because more or less he copied her future." But María took a long view: "Once he finishes high school, maybe he will study something else; he's just more immature [than his sister] about his ideas."

In this family, we have an account of two siblings with very similar (indeed, co-developed) skills on their XO laptops—skills that were a product of their own interests and, in the spirit of connected learning, scaffolded by their mother toward educational ends. Of course, they did not develop completely identically: Elisa's interest in Japanese, art, and engineering expanded her uses of her XO and later her Acer laptop, while Manuelo was less sure of his identity and interests aside from the computer—despite the efforts of Paraguay Educa to shape them. The fact that, by 2013, Elisa had given up the identity of *Scratchera*, which had given her such pride in 2010, likely speaks to the unequal opportunities that Paraguay Educa provided to these two siblings: it tried hard to cultivate the *Scratchero* identity in Manuelo but put less work into Elisa. Even so, as he entered high school, he "copied" the career path of his sister rather than adopting the imaginary of the technically precocious boy as part of his own identity—an imaginary that he never expressed familiarity with.

What would these youth have done if they had not been part of Paraguay Educa's OLPC project? It is impossible to say for sure; María's sustained engagement with Paraguay Educa shaped her views of the role that computers could play in children's lives. But the focus she already had on education and critical thinking, the presence of a computer in their house, and the fact that the eldest daughter had already attended college suggests that they were already within a community that would likely have led to higher education and a professional job of some sort—even if it was circumscribed by the Caacupé area rather than connected to the cosmopolitan identities and communities that Paraguay Educa's Asunción-based staff, and many of OLPC's tech-elite contributors, were accustomed to. And it appears that this community, rooted in family ties and values, remained the strongest of the various social worlds in which these siblings lived and learned.

Isabel the *eToysera* and Her Friend Nelson

Other youth were not as celebrated by Paraguay Educa as Manuelo but were just as interesting. In 2010, one such student especially stood out: Isabel. When I first met her, Isabel was a sassy and outspoken sixth-grade student in a large private school in downtown Caacupé, a few blocks from Manuelo and Elisa's public school. Isabel had lived with her aunt, grandparents, and uncle since kindergarten, a living arrangement that was not uncommon in Paraguay, where single teenage mothers often had to lean on their families for help with child-rearing while they worked long hours, sometimes in other cities. Isabel helped her extended family with their business selling the Paraguayan staple *chipa*, a hard corn biscuit fashioned into a ring. Riffing on the term *Scratcheros* that she heard from Paraguay Educa staff and trainers, Isabel called herself an *eToysera*, a lover of the program eToys.

When I first got to know her during my 2010 fieldwork, Isabel was as loquacious as Manuelo and Elisa were reserved, and she had many opinions about the XO and the laptop program. Right after introductions in our first interview, she launched into a rapid-fire description of her laptop use: "I love using my XO. There are so many activities I like, especially eToys. I use it here and in school. Sometimes the teacher wants to do a project in Turtle Art, and while she is still explaining, I've opened eToys and have finished the whole project there." She went on to explain that she regularly did her homework in the program that her teacher or the school's trainer (employed by Paraguay Educa to encourage laptop use) asked the class to use, such as Scratch, and then completed it again in eToys and showed her teacher both. She also liked to teach her teacher and peers new features on the laptop, finish longer-term projects the day they were assigned, collect information off the internet for her teacher to use in upcoming lessons, and even take over lessons when the teacher had to leave the room, earning herself the nickname "little teacher." On helping others use the laptop, she said, "They [my friends] will ask me how to do their homework, and I show them. Some get frustrated because they don't understand, but I just show them again. ... I will do a project at home and then go show [my teacher] and describe it to her. I helped my teacher a lot last year, and this year also."

Isabel outlined her laptop practices specifically in opposition to those of her classmates. "My friends search for games. I don't search for them," she explained to me. She went on to describe trying to work on a project with a friend who started up a video game instead of searching for the information the teacher had suggested. "I told him, 'What great information you have,'" she said, laughing. She continued, "I don't download games. I search for information, and that's all. [My teacher] says I am very mature for my age." She said that she had developed her distaste for games and television ("Boring!" was her assessment) in conversations with her caretaker aunt, who was immensely proud of Isabel, and in opposition to her uncle, who was only two years older than her and spent a lot of his time watching television, playing video games, and avoiding schoolwork. She did play a car-racing game with her uncle for a while, but only "until I started beating him—then he didn't let me play anymore." She was otherwise actively disinterested in games and had none installed on her XO.

Most of Isabel's time on the laptop was spent either going above and beyond on her schoolwork or searching for information to support classroom activities. Because she was not very interested in games, she did not feel motivated to make them. As such, she did not engage in the kinds of programming activities that had sparked Paraguay Educa's interest in Manuelo. Thus, though Isabel was beloved by her teacher, teacher trainer, peers, family, and even me after I met her, and though she arguably gave back more of her knowledge of the laptop to those around her than the shyer Manuelo or Elisa, she was not recognized by those who could provide opportunities for her.

Isabel was not able to show me most of her projects during her first interview because they had been erased in a software upgrade, which other youth had described as well. "I lost all of my work when we upgraded Sugar," she lamented, frowning, her brows knitted. "I was so sad, because I had all of my work there; it took me five hours to finish an assignment about a novel. They didn't tell me everything would be deleted, or I would have saved the information on a pen drive [USB drive]." However, unlike some of her peers, this did not demotivate her from using her laptop. The troublesome trackpad did not dissuade her either; her aunt saved up and bought her an external mouse instead.

Isabel took fastidious care of her XO laptop, washing her hands before using it and cleaning it daily with alcohol wipes and cotton swabs. She expressed disdain for classmates who broke their laptops and did not fix them. She was also not shy about telling others how to care for their laptops. "For parents' day [at school]," she explained, "I had the idea to teach a class for parents about how to clean and take care of the laptops. Nelson [my best friend] will help; he will make a presentation in Scratch, and I will do one in eToys." She was also not shy about giving suggestions for improving the program. She especially wanted video chat and asked when she could follow up with me to see if it was done. She was so enthusiastic and insistent that I didn't have the heart to tell her that OLPC developers had told me that the XO laptop had been specifically designed to not allow video chat for fear of sexual exploitation.[15] In our second interview in 2010, she said that my website should also be available in Spanish, shaming me into adding a Google Translate widget that very evening.

In her 2010 interviews, Isabel often mentioned her friend Nelson, and Nelson likewise talked about Isabel. Nelson liked to use Scratch instead of eToys, but the two would explore new things together or work in parallel on projects in their respective programs of choice. Nelson also liked to keep on top of software upgrades, getting the latest installation files from one of the teacher trainers and distributing it to friends, though this was responsible for Isabel losing her projects.

Nelson was influenced by Isabel to explore Scratch—but he was also interested in the media that other friends and classmates were. Nelson used Scratch when he spent time with Isabel, but it was he whom Isabel chastised for playing a game instead of working on a project in class, and he said that most of his laptop use was for games and music. "Best are the three [games] from Super Nintendo," he told me in an interview. Nelson also used his laptop to watch movies. "I have a movie on my pen drive from a friend. I usually like comedy or action movies," he said. He went on to explain how his habits have changed as a result of the XO: "I don't see actual movies or watch TV now," he said. However, he also read less than he used to, instead playing games or using Scratch on his laptop. "I like to read; I used to read a lot, but I don't much anymore," he told me. Still, Nelson used his laptop more than his brother, who was a nonuser: "My brother also has an XO, but he doesn't like using it as much. He would rather play other things."

Isabel's aunt, who had raised the girl since she was five, told me in a 2010 interview that she discussed television watching with Isabel, and they agreed that it was not a good use of time, though these discussions did not seem to have as much of an effect on Isabel's teenage uncle. But she chalked it up to personality: "She's always like this with everyone, all enthusiasm, showing them every little thing." She was so happy that Isabel had received her own computer. The family—like Manuelo and Elisa's—were among the minority of Caacupé residents who had a desktop computer at home before 2009, but she explained that when Isabel had used it frequently, her uncle or brother would sometimes go on and delete all of her files. But the XO changed all that. "This way, she can have her own," her aunt explained.

In 2013, I encountered a much-changed Isabel. Her aunt had moved to Spain in 2011, and Isabel now lived with her father and grandmother in a small rural house. Rather than inviting me in, she brought two chairs out to the dirt driveway for the interview. Her sassy openness had been replaced with guarded wariness. She told me that she had given her XO to her younger brother in fifth grade after he broke his own. Instead, she used a desktop computer, and although she had Scratch and eToys installed on it, she said that she had not done much programming in the previous couple of years because she no longer saw much of her beloved teacher or of Paraguay Educa's erstwhile trainers. She had also lost touch with Nelson, who had moved to Asunción, and she did not have other friends interested in exploring Scratch or eToys with her.

Instead, she spent a lot of time playing video games—particularly Minecraft and an online multiplayer game called DOTA 2 (the second Defense of the Ancients, a multiplayer online battle arena game). She even entered in regional eSports video-game tournaments, competing in DOTA 2 for small prizes of cash or computer equipment. She told me that she wanted to study graphic design at Caacupé Technical College starting the following year and then to be a video-game artist—but she also wanted to be a veterinarian. More than anything, she wanted to travel: to Spain to visit her aunt, to the United States, and around the world to participate in international DOTA 2 eSports tournaments. She said she would love to learn English but had not had the opportunity to do so.

In Manuelo's, Elisa's, and Isabel's stories, we see some similarities: bright and driven youth, shaped in part by their own interests and in part by their environments to explore some of the more advanced creative tools on their XO laptops. Similarly, their respective caretakers took an active role in shaping their lives and interests, especially when contrasted to the less active role that many other parents took toward their children's computer use or leisure time—which shows that their role in encouraging creative laptop use was significant. This problematizes the exclusive focus on child-laptop interactions that OLPC tended to favor in lieu of a more holistic, social approach to examining learning behaviors, especially the influence of peers and parents.

But there was a notable difference between Isabel and the Scratcher siblings. While Manuelo and Elisa's mother, María, kept in touch with Paraguay Educa and made sure that the NGO knew of her children's accomplishments, Isabel lacked such an advocate, and her interests developed within a different community, influenced more by her gamer uncle than by Paraguay Educa. This disconnection had material effects on Isabel's life: even though she wanted to learn English, she was not one of the Caacupé residents to whom Paraguay Educa offered four years of English lessons; even though she wanted to travel, she was not chosen to visit MIT; and even though she was passionate about eToys, her identity as an *eToysera* was not recognized or reinforced. Much like Elisa, eventually her interests shifted.

I do not mean to censure the Asunción-based employees of Paraguay Educa for ignoring seemingly talented youth such as Isabel. Given the relatively infrequent visits by Paraguay Educa's leadership, it is understandable that the children of parents connected to the NGO, such as María's children and those of Paraguay Educa employees in Caacupé, would catch their attention the most. I do suspect, however, that in addition to lacking a well-connected advocate like María, Isabel's 2010 use, as focused as it was on schoolwork, may not have fit the social imaginary of the technically precocious boy that Paraguay Educa appears to have been looking for among these youth. The gendered expectations around what uses were considered interesting and who was expected to exhibit them thus did not work in her favor.

Moreover, this disconnection shows that the worlds these youth found themselves in, and what they made of them, had perhaps more to do with chance than with skill or promise. Paraguay Educa's code clubs, English classes, and trips to MIT were attempts to structure the interests and even the identities of these youth, to create a different kind of community among those identified as most promising, given limited resources—in effect, to produce and enforce the charisma that had inspired them to believe in the project. But the process of being recognized was biased. Moreover, even among those youth who were identified as talented, such as Manuelo, there were other social norms and expectations that at times competed with Paraguay Educa's aims to instill a culture that valued programming. These social norms may have been even stronger among youth who, like Isabel, had more tenuous connections to Paraguay Educa. In this way, the translation of the XO laptop's charisma from One Laptop per Child to Paraguay Educa to children in Caacupé was incomplete, and its charismatic promises were unfulfilled even among these highly-engaged youth. Rather than transforming their social worlds, these youth remained embedded in them, still beholden to the structural limitations of gender, place, and social class.

Precocious Youth, Supportive Parents, and Environment

Manuelo, Elisa, Isabel, and Nelson were fairly unique at their respective schools, but in 2010, I found multiple students engaging in more creative pursuits on their laptops at a parochial school called Futuro, which was sponsored by an evangelical church in Texas. Futuro followed an American school schedule (seven in the morning to three in the afternoon) and used (evangelically themed) textbooks for lessons—a contrast to other Caacupé schools, which ran two four-hour school days for two different student bodies and generally had few or no textbooks. English lessons were also part of the curriculum, though the level was very rudimentary: Futuro students would manage a "Hello!" and perhaps a "How are you?" with me before slipping back into Spanish. While students did not have to pay to go to Futuro, they did have to score well on an entrance exam—a practice that favors those with the means to prepare. As a result, Futuro attracted many of Caacupé's more privileged children, whose families had the time and resources to not only have their children take the test but to help them score well.

Almost all of the families of Futuro students who used their XOs more creatively, like Manuelo's, Elisa's, and Isabel's, already had computers at home when they received an XO laptop from Paraguay Educa—though this was a rarity in Caacupé more generally. Moreover, many of them had caretakers, like María and like Isabel's aunt, who encouraged critical thought regarding media consumption. Thus, reflective computer use was already a part of their lives, and the XO was more likely to fit into those existing practices. For instance, Rosa, the mother of a set of twins in sixth grade who were trying to recreate the Mario Bros. game one scene at a time in Scratch on their home computer, had told her children that she did not want them spending their time playing video games at home—but they were allowed to *make* them. An articulate former community organizer who made a living as a seamstress, Rosa explained that though she did not have time to learn about computers herself, she still monitored her children's usage and would have them teach her how to use the laptop so she could stay on top of what they were doing. "I observe a positive change in them" as a result of the program, she told me. "They're more creative; they're always creating things and are curious and want to investigate, and to me this is very good. I like it when they're curious." At the same time, she did not want her children to get too wrapped up in the computer. An avid environmentalist, she often mentioned wanting to cultivate in her children a connection with nature, something she thought that too many Paraguayans lacked. "I think the world of the computer is constructive and educational, but I believe in contact with nature, in going outside to play, getting dirty. ... They need to know there are other important things in life. Computers aren't negative, but I want to give them many options."

Rosa also described the tension between guiding her children and letting them develop on their own, echoing María's words. "What's most important is teaching your child. Parents can show them what's good and bad, tell them what's right and wrong, and open their minds so they can tell themselves," she explained to me. "I give them a lot of freedom to choose. My kids belong to society and themselves and will have to find their own way in the future." Later, she told me that even though her nascent career as an ecologist was cut short by an early pregnancy (common in a country where birth control was very difficult to access and abortions were illegal), she loved teaching her children about what she cared about in the world and encouraged them to be just as passionate: "Working with children is a project with a lot of potential. That's where it all begins, in childhood. When the child grows up, they act based on everything in their heads, and how you did teaching them can be assessed then."

As I got to know María, Rosa, and the other parents and caretakers of precocious youth in Caacupé, I found that even prior to the laptop, they all had steered their children toward creative and critical thinking. As a result, these youth were often among the best students in school, and their parents or caretakers (largely mothers, who were solely responsible for child-rearing in many of the Paraguayan families I observed) had already discouraged them from watching television in place of reading, drawing, playing, or similar activities. In interviews, their mothers—or in Isabel's case, her aunt—spoke articulately about what they considered educational and their goals of fostering creativity in their children. Even though not all were adept at using computers themselves, they steered children away from uses that looked a lot like television watching and toward more creative activities. Among those I observed in Paraguay, all cases of creative exploration were coproducts of youth, parents, teachers, and other influences. Thus, even though Paraguay Educa lauded some of these youth—especially boys such as Manuelo—as natural geniuses, this suggests that their interests were not innate but were influenced by their social worlds.

There is also a clear link in these accounts to connected learning—learning that begins with youth's own interests, even if those interests are media-centric, and scaffolds them into something more "educational." Even the XO uses I saw among these precocious youth often blurred the line between media machine and learning machine. At school in particular, many of them were interested in the same kinds of media-consumptive practices that their friends and classmates were, though a few, such as Isabel, formed oppositional identities specifically reacting to them.

What does it mean for One Laptop per Child that even in a well-supported project such as Paraguay Educa (at least in its first years), so few children engaged with their laptops in the way that OLPC had hoped? As impressive as these youth were, it is striking that so few used their laptops in the ways that OLPC literature and publicity materials had framed as both natural and inevitable. This calls into question the universality of the social imaginary of the technically precocious child (often portrayed as a boy) who uses a natural technical creativity and rebellious inclinations to develop skills without the help of (or even in spite of) adults, leapfrogging past them in abilities. Instead, children's motivations appeared to be less the products of individual interactions between the natural, supracultural child and the Protean laptop than a negotiation between many factors, especially parents and peers. In short, laptop uses and meanings were socially created and socially mediated—and, at times, operated under different cultural logics in Caacupé, where the imaginary of the technically precocious boy did not seem to be legible. Thus, though Paraguay Educa arranged for those they saw as the most precocious to attend coding clubs and English lessons, these youth still mostly learned within the social world of Caacupé, and their futures likely did not change radically from what they would have been without the laptop project.

The observation here that learning is a profoundly social process is nothing new to education or sociology; the idea is well established, both in and out of the classroom.[16] Yet educational projects such as OLPC, and more recently the resurgence of interest in self-directed online learning, tend to ignore this evidence in favor of the much more charismatic story that learning motivation is something that can be fostered in the relationship (such as it is) between the student and the machine. As a result, these projects can lack a grounding in social or institutional support for creating environments where such learning could take place, and even their success stories risk entrenching the social divisions they seek to erase.

Language, Marginalization, and the Cruel Optimism of Programming Competitions

There has been one undercurrent throughout this chapter that I have not yet unpacked: the focus that many of the most creatively engaged youth had on learning English. Paraguay Educa had arranged for Manuelo, Elisa, María, and twenty-two other Caacupé residents to receive four years of English lessons from the Paraguayan-American Cultural Center in town, lessons that were otherwise expensive. The students of the evangelical-sponsored school Futuro also learned a small amount of English. Some from outside those worlds, such as Isabel, also wanted to learn English: she explained that she wanted to understand all of the search results as well as the parts of Scratch, eToys, and the Sugar interface that had not been translated. She told me in 2010, "I use Navegar [Browse] a lot. My friends and I downloaded Mozilla, and we use Firefox more. It's all in Spanish, and the other is still in English—things in Navegar—and for this we use Firefox to search."

Elisa liked to translate English song lyrics into Spanish, and she also used Google Translate on English search results when Spanish search results turned up empty. She explained, "When I'm wrong about something, I look for a solution. I'll type this into Google Translate and switch to English and search. ... For example, I wanted to color something black, but [the drawing program] only had [color] names in English, so I copied the translated name. ... Or for the operators, I don't know a lot about them so I'll get help, but it's in English." I then asked her if she wanted to learn English, and she replied, "Yes, so I know more what it says." Her brother Manuelo likewise had needed to learn some English to be able to program in Python, one of the emphases of the XO Evolution club, even though he did not often attend. Their mother, María, thought that English would help her children as they transitioned into the professional careers she hoped they would have. "Why English? First, because it is one of the universal languages," she explained to me in a 2013 interview. "[Manuelo] needed a lot of English when he was in the Python part. ... I mean, he was stuck with some question in the programming part, and that part is all English, and he had a hard time not knowing, right? Then he revived and used Google Translator, but it is not the same because words might mean other things."

English is not one of Paraguay's two official languages; those are Spanish and Guaraní. In theory, everybody learns both nationwide, but in practice, many rural residents speak mostly Guaraní whereas those in urban centers, especially Asunción but also within downtown Caacupé and other provincial towns, speak mostly Spanish. The XO laptop's Sugar interface was, in theory, translated into Spanish, but as Isabel pointed out, the translation was incomplete. Moreover, youth were constantly encountering English-language search results.

The necessity to learn English to code—especially as participants in Paraguay Educa's Saturday clubs switched from the block-based programming in Scratch to text-based programming in a language such as Python—was driven home to me in watching the XO Evolution club prepare for the finals of a programming competition during my November 2013 fieldwork visit. This club, founded around the time the teacher-trainer program was discontinued, occasionally drew Manuelo and Elisa in, but during the weeks I visited in 2013, it consisted of a group of eight high-school boys whom I had observed only in passing in 2010. It met in the downsized Caacupé headquarters of Paraguay Educa, where the NGO's skeleton repair crew stored hundreds of broken laptops that were waiting (perhaps forever) for repair parts. The boys hunched amid these electronic husks where, guided by two of the former teacher trainers, they worked to learn Python well enough to be able to solve the technical questions typical of programming competitions: take in a text file of input, parse it, process it (often with the clever use of a standard computer-science algorithm), and output the results in a very specific format that could be checked by an automated script.

This preparation was an uphill climb for them all. They were focusing on learning Python, one of the three languages in use for the competition and by far the easiest—it was a language designed for learning, easier than the other two choices of Java (an object-oriented language that tends to be very verbose) or PHP (a scripting language that tends to be both more opaque and fussier about its syntax than Python). Even so, all of Python's commands and error messages were in English, which none of the boys knew well, even though some of them had taken several years of the English classes that Paraguay Educa had arranged. The trainers were not much better off. As a result, my role as ethnographer blurred into mentor as I helped them make sense of the error messages that were generated by simple misspellings, such as "pirnt" instead of "print" or "inpu" instead of "input." They were trying hard but had few resources to help them scaffold their Spanish- and Guaraní-language skills into the English-centric demands of text-based programming.

The day of the competition, the boys met at four in the morning in central Caacupé to take public buses to the capital, but even with the resulting lack of sleep they were ebullient. I met them and the trainers outside the private college in the suburbs of Asunción where the event was held, in a fairly new building set back from the honks and yells of the bustling street. They talked excitedly about how they would leverage their first-prize trip to California into jobs in the United States and prospects for their families. Soon the eight boys—working as two teams of four—joined some forty-odd other competitors in the college's computer lab. Everyone else waited outside while over a dozen teams of two to four students all around the room huddled together, whispering, in front of the rows of black workstations.

While we were waiting, I talked quietly with other mentors and event staff. I learned that the two teams from Caacupé were the only participants from public schools and the only ones from outside Asunción. Other students, largely coming from wealthy families, generally spoke good English—a priority even over Guaraní in Asunción's wealthiest private schools—and some had been to this competition repeatedly. The school that sent the most teams, which was not far from the site of the competition, had also won the competition for several years running. Many of its graduates went on to attend the best university in Paraguay or a university abroad. This was the cosmopolitan world in which the employees of Paraguay Educa had been educated—a world with far fewer of the structural constraints that Caacupé residents often experienced.

After three hours of nervous whispers and anxious typing in the competition room, the judges called time to a collective groan and burst of discussion. The team mentors and I joined the competitors, and we ascended from the second-story computer lab to a fourth-story room overlooking the flat, red-dirt and green-tree expanse of Asunción, with the Paraguay River and Argentinian countryside in the distance, to await the judges' determination. The November day was heating up—at eleven in the morning, it was already a humid 35°C (95°F) outside—and the one air-conditioning unit in the back of the room worked hard to overcome the heat leaking in through the louvered windows and cinderblock walls.

The organizers had lined up several speakers to give talks about programming and Paraguayan entrepreneurship while the judges determined the winners. While we fanned ourselves and waited, we heard from the winner of a regional app-development contest, who wanted to address the lack of municipal services, such as garbage collection, in Paraguay. He celebrated the submissions that Paraguayan programmers had made to regional app contests but concluded with advice in a quite different tone: the most important thing for learning to program was to learn English. Another speaker took the stage and spoke at length about how English was a common point of understanding between Brazilian and Argentinian hackers and the substantial US expatriate hacker community. Assuming that all in the room were from elite backgrounds like they were, these speakers spoke of leveraging private-school educations into strong SAT scores and SAT scores into college in the United States or Europe. Nobody mentioned the many barriers that might be in the way. Likewise, nobody noted that, of over fifty competitors, only three were girls. Indeed, the two teams from Caacupé both consisted entirely of boys.

During these talks, the Caacupé team members were fidgeting and whispering to one another, only partially listening in their excitement to hear the results. But their victory was not to be: a team from the private school that had won in several previous years prevailed again. Crestfallen, the Caacupé teams nevertheless posed for a photo next to the competition banners after the winners took theirs. We provided lunch at the local cantina, and they boarded buses that would take them back home.

My experience watching these boys prepare for and compete in this event highlights many of the structural differences—and the points of active marginalization—between the Caacupé-based XO Evolution computer club (even sponsored as it was by Paraguay Educa) and the elite teams based in Asunción. There was the physical distance: the competition was of course in Asunción, not Caacupé, which marginalized potential participants from outside the wealthy capital. There was social marginalization, where competitors were assumed to be wealthy, attending the best private schools, and able to afford the best colleges in the country or even the world. The lack of girls in the room demonstrated a gender marginalization that seemed to be completely taken for granted by those present. And, finally, there was a linguistic marginalization, in which knowledge of English, not Spanish or Guaraní, was both necessary for understanding the programming languages used in the competition and emphasized by the speakers as necessary for entering the world of computer programming more generally.

What can we make of this focus on English? OLPC founder Nicholas Negroponte has described children's interest in learning English as good and natural. In his public talks, he has referred to English as the lingua franca of the internet and its acquisition as necessary for global citizenship. "As children grow up on the internet they're going to be more global than their parents," he explained at the World Economic Forum in 2006. "They're going to at least speak two languages. They're going to speak their own language plus English. We are definitely going to have a global culture, and we're going to have a single language." He went on, expressing shock that kids in Switzerland, where the forum was hosted, did not all speak English, and he derided the "preservation of cultures, which usually means the preservation of a particular language—I sit and wonder to myself, 'Is that really necessary?'"[17]

At least in the programming world, this perception is commonplace. In a study of the Lua programming language's Brazilian developers, for instance, Yuri Takhteyev found English to be the norm even when programmers were talking to others within the Brazilian programming community. Moreover, rather than developing Lua to be Portuguese from the bottom up, these programmers adopted the conventions of the international programming community—conventions that Negroponte alludes to—and made the programming language English in order to make Lua legible and usable worldwide. Takhteyev's participants were generally more cosmopolitan than the youth in Caacupé whom I followed; they were able to connect to what Takhteyev has called "networks of practice," which included expectations and resources that helped them learn English well enough to program computers in the language (a variation of Jean Lave and Étienne Wenger's concept of "communities of practice," extended to include not only people who scaffold learning through an apprentice-like model but documents, software, and other nonhuman resources that also play important roles in the learning process).[18] From this perspective, Negroponte's stance makes more sense, especially because, as previous chapters have demonstrated, one of OLPC's implicit goals was to make it possible for all of the technically precocious boys of the world to join the hacker community.

From the frame of contemporary cultural anthropology, such an argument is much more problematic: the opinion that English should be the universal language (and, often as an implied corollary, that US or European culture should be the universal culture) is clearly paternalistic. At the same time, Negroponte was perhaps on to something when he said that efforts led by people in the Global North to preserve "traditional" cultures can also be paternalistic (though burgeoning movements of cultures to revive themselves, pushing back against quite recent efforts to eradicate them, are very different). As anthropologist Eric Wolf has shown in *Europe and the People without History*, faulty assumptions set up "the West" as a social imaginary that is allowed to have a history and is assumed to be naturally dominant—in opposition to "local" cultures, which are assumed to be both timeless and fragile to the onslaught of Western norms. Non-Eurocentric histories, much less the complicated interconnections between all of these cultures, tend to be ignored in this narrative of timeless and provincial indigeneity.

Yet, even as Negroponte has decried protecting cultures for their own sake, the imagery that One Laptop per Child often used to promote its laptops—children using or carrying XO laptops in pastoral settings or "traditional"-looking huts—evoked this timelessness. As Anita Say Chan has also noted, from the cover of a book by Walter Bender and colleagues (showing two children holding XO laptops by the handles while crossing a stream) to Negroponte's OLPC presentations (showing children hunched over computers in fields and forests) to the imagery used by individual projects in Peru and Uruguay, laptops have been inserted into what appear to be timeless natural settings, evoking a narrative that erases the thorny marginalization that this chapter has demonstrated in favor of the simpler and much more charismatic narrative of easy, child-led cultural uplift through their "natural" technical curiosity.[19] Thus, just as the British Empire did in Wolf's account, One Laptop per Child at once asserted its assumed natural cultural dominance ("Of course children want to learn English and learn to program!") and erased the historical interconnections and ongoing subjugations that had put Europe and European colonizers in a position of material power over much of the rest of the world.

Both approaches fail to account for the material realities and the dreams of those whose cultures, languages, and futures are at stake. Anthropologist James Ferguson brings this into sharp relief in a story from his first fieldwork in Lesotho, told in the introduction of his book *Remotely Global*. He recounts extolling the insulative and locally sourced virtues of the traditional thatch-roofed, one-room houses over the new "European-style" tin-roofed houses to one of his participants, only to have the participant retort, "How many rooms does *your* house have?" Ferguson realized that the participant's aspiration to build a tin-roofed house "was not a matter of blind copying; it was a powerful claim to a chance for transformed conditions of life." Ferguson continues,

> The connection of cultural difference to social inequality is a theme that is insufficiently appreciated in much recent thinking about what is sometimes called global culture. For cultural practices are not just a matter of flow and diffusion or of consumer choices made by individuals. Instead, they index membership in different and unequal social groups, globally as well as locally. In this sense, yearnings for cultural convergence with an imagined global standard ... can mark not simply mental colonization or capitulation to cultural imperialism, but an aspiration to overcome categorical subordination.[20]

For the youth discussed in this chapter, language represented a site of yearning as well as an index of marginalization. The primacy of English on the internet and in programming languages represented to many youth and families the possibility of a cosmopolitan future—a future that the XO laptop, the English-heavy internet, contact with the Asunción elite, and contact with international visitors reinforced.[21] At the same time, the barriers to learning English reinscribed the peripherality of Caacupé and actively marginalized its residents, in the same way that the slow buses between Asunción and Caacupé physically separated those outside the capital city from sites of state power. It is not that Isabel will never travel or the code club will never visit California—but such outcomes are less likely given the material realities of these youth and their families. Although it is no more my place than it is Nicholas Negroponte's to celebrate one culture over another or to point out to these youth the potential limitations of their dreams, it is worth attending to the consequences of what cultural anthropologist Arjun Appadurai calls "an ironic compromise between what they could imagine and what social life will permit"—the kinds of difficult-to-change structural differences that Lauren Berlant references when she refers to the optimism that might result from these visions as "cruel."[22]

This was especially salient among those members of XO Evolution who attended that November 2013 programming competition. This group of high-school students—notably all boys, at least during the weeks I attended—had self-selected into the club and, with Paraguay Educa's influence, had become at least somewhat familiar with the hacker identity that it sought to promote. They had taken on a version of this identity as their own, though they pinned onto this imaginary less the specific politics of free software or the hacker ethic—which were largely outside of their experiences and concerns—and more a pathway to travel, financial security, or social change. In my observations and interviews with them, I learned that they found contact with Asunción's elite and with international visitors like me exciting—it made them feel important and powerful. But unlike the Brazilian Lua programmers whom Takhteyev studied, the national context and the local communities they grew up in made it more difficult to achieve these interests and aspirations.

Paraguay Educa was trying to provide an alternate social world and set of social imaginaries that could potentially scaffold these boys into the world of programming. But the individualist messaging of both One Laptop per Child and Paraguay Educa was counterproductive to this aim. Taking a page from the American "pull yourself up by the bootstraps" self-made-man mythology, both organizations portrayed hard work as the key to success, ignoring the intensely hard labor that so many across Paraguay—from subsistence farmers to schoolteachers—engaged in every day. Meanwhile, the structural and social forces that continued to push these kinds of communities to the periphery went unacknowledged and unexamined. The hacker identity that resulted was one that was still deeply influenced by privilege, infrastructure, and proximity to cosmopolitan cores and international networks. Even though anthropology has deconstructed notions of center and periphery, the technological elite in Paraguay reinscribed their centrality and the peripherality of those they sought to help.

6 Performing Development

The peak of the hype cycle was ... obviously deeply ideological, and interesting as
such. But it also registered something real about the performative efficacy of what
one might call "technological charisma." That is to say, the hype was not just
empty; rather, it brought about its own concrete social effects "on the ground."
—William Mazzarella, "Beautiful Balloon"

For three days in the middle of October 2010, almost four months into my
fieldwork, one of the founding members of One Laptop per Child and its
former vice president of content, Walter Bender, visited the OLPC project
in Paraguay. Paraguay Educa, the NGO running the project, had carefully
planned his itinerary, circulating it to staff and volunteers (including me)
via email before his visit. On his first day in town, Bender met with government officials whom Paraguay Educa hoped would support and expand the
project going forward. That evening, he gave a keynote talk at the Latin
American Informatics conference, where he extolled the virtues of Turtle
Art, Sugar, and OLPC to a full university auditorium. His third and last day
in town was entirely taken up by a visit to Itaipú Technology Park, one of
Paraguay Educa's principal funders and the operator of the Itaipú Dam,
the world's second-largest hydroelectric dam, on the Paraguay-Brazil bor-
der. For the morning of Bender's second day, Paraguay Educa had planned
a "tour of schools" (*recorrido de escuelas*) in Caacupé to observe the project
firsthand—and I was intrigued.

At this point in my fieldwork, it was clear to me that OLPC's vision of
an effortless transfer of mathematical knowledge via a charismatic machine
to children passionate about learning how to program was, to say the least,
considerably more complicated in reality. The previous three chapters have
demonstrated these complications, showing that the social imaginaries

upon which OLPC's charisma depended did not find the purchase among Caacupé residents that Paraguay Educa's founders had hoped they would. In fact, the project's results were similar to other one-to-one laptop projects around the world, despite the charismatic machine and history-defying rhetoric of OLPC. As such, it could hope to achieve at most what the most well-supported one-to-one programs achieve: perhaps some incremental improvements in reading and probably increased technological familiarity, with more benefits likely going to the children who started out with more privilege.[1]

What would Bender make of this rather uncharismatic reality? Would he see some of the design flaws that made the laptop harder to use and make sure that they were fixed in future hardware and software iterations? Would he revise his assumption that kids would fix their own laptops—which he had repeated just the night before in his Latin American Informatics conference keynote—or arrange for OLPC to better support projects such as Paraguay Educa's? Would he revise the broader assumptions behind this thinking: that little was needed for the project to be successful other than laptops and children to use them and that these children (or at least the "most intellectually interesting" among them) would naturally take interest in the machine and spontaneously, easily, and joyfully bring about the cultural change that OLPC hoped to see?[2]

The day started like many school days in Paraguay. It was a Tuesday in late October 2010, well into spring. The sun was beating down, and temperatures were already approaching 30°C (86°F) by eight in the morning, when two Paraguay Educa employees, Bender, and I parked outside the school, just in time for the start of classes.[3] We had carpooled together to Caacupé from Asunción, rising before six in the morning to make the slow journey. There had been some question whether there would even be classes today, but the teachers' unions decided to keep their strikes to Thursdays, so school had convened—and here we were. We met six of Paraguay Educa's fifteen teacher trainers outside the school, as well as the local project coordinator.

We all assembled at the edge of the school's courtyard, where children were mingling, some playing pickup soccer games. Bender took out his blue XO laptop—designed for high-school students in Uruguay, with a larger keyboard than the standard green XO had—and showed it off to any children who wandered by. The trainers checked in with the director and some teachers at the school. Shortly after, the director emerged from the front office with a large brass bell and rang it overhead. Around one hundred students in their uniforms of white shirts and dark pants or skirts gathered under the eaves of the school building or in the shade of trees at the edges of the courtyard and turned their attention toward her. She led them in a school-wide morning song and then launched into a speech. "Students," she said in Spanish, "listen to everything this man has to say. He is a visitor from the United States and came all the way here to help you discover new things." The children clapped. Bender, who did not speak Spanish, stood nearby, smiling blankly.

She continued, "This is a huge opportunity for a special group of you. If you are in the Turtle club, come up here. These students are going to participate." Seventeen children detached from the crowd and gathered where the director had indicated. Nearly all of them were wearing uniforms for different schools or everyday clothes, indicating that they did not attend this school. I recognized them as students who were part of one of Paraguay Educa's Saturday clubs, the one focused on Turtle Art. A recent implementation of Seymour Papert's Logo environment (though borrowing the block-based programming paradigm of Scratch), Turtle Art was the Sugar activity in which Bender was most invested, as he had described in detail at his keynote the night before. Teacher trainers had reached out to members of this club to attend today, and they were missing a full day of classes at their own schools to be here. While Paraguay Educa and I knew about the labor behind assembling this group of students, I was not aware of anybody mentioning it to Bender, and he gave no indication of noticing the discrepancies in uniforms.

The group posed for photographs while the rest of the school looked on, then around 8:30 a.m., they were released to their classrooms while we were ushered into a spare classroom next to the director's office. Bender, standing behind the teacher's desk at the front of the room, plugged his laptop into a projector that the trainers had set up and launched into a Turtle Art demonstration. He discussed a few advanced math "tricks" in Turtle Art, including variations of Papert's "circle" algorithm, where to make a circle one can move forward a bit then turn a bit, many times until back to where one started.[4] As he stepped through some examples, the two Paraguay Educa employees from Asunción stood up front with him and translated what he said into Spanish. The six teacher trainers, Paraguay

Educa's coordinator, the school director, a teacher, and I sat among the students or stood at the back of the room. I spent the hour busily jotting down field notes: who was there, where they were in the room, what they were doing, what was said, and what else transpired. Later, when I analyzed my data, I realized that with seventeen children and thirteen adults in the room, the ratio of children to adults in the room was less than two to one—whereas in a typical classroom, it could be as high as twenty-five to one. Some of the children followed the director's advice and listened to Bender, with Turtle Art open but untouched on their laptops, but from the back of the room, I could see that a few had opened web browsers on their laptops instead.

After an hour of talking, Bender closed his laptop, and we packed up and left amid children streaming out of other classrooms for recess. Though the itinerary had called for a "tour of schools" plural, this was the only school visit we did that day. Our next stop was a café above a supermarket on the edge of town, where we ate an early lunch with the coordinator and a few of the trainers. There, the conversation revolved around using Turtle Art as a solution to any need, from teaching mathematics to conference presentations, as Bender had done for his keynote presentation the night before (though with some technical difficulties, even for the MIT research scientist and skilled programmer).

We also talked about an abacus variation that was developed for the XO laptop by one of Paraguay Educa's trainers. Bender had misidentified the trainer as a "local teacher" in his keynote the night before and had implied that their innovation was spontaneous, rather than it being their job to find ways to use the laptop in classrooms. Paraguay Educa employees and I tried to correct this misconception by explaining the difference between teachers and trainers. Still, the same mistake was repeated in the book on OLPC that Bender coauthored and published some two years later. "Teachers in Caacupé, Paraguay," the book states, "were searching for a way to help their students with the concept of fractions. After playing with the Sugar abacus activity, they conceived and created a new abacus that lets children add and subtract fractions. Sugar didn't just enable them to invent, it encouraged them to invent."[5] This book, *Learning to Change the World*, cited the classroom visit I had just witnessed as proof that the project was working: children were engaging with their XO laptops; they were excited about Scratch, Turtle Art, and the other constructionist activities on the machine; and they were going to change the world with this knowledge (as the title itself states).

I read *Learning to Change the World* as I was finishing my dissertation in early 2013 and was perplexed that the book could make such claims, based on my experience of the project and the parts of Bender's 2010 visit that I had witnessed. While he did visit an actual school, with actual students (though mostly not students from that school), the whole encounter was so scripted that it was not possible for him to get from that visit a sense of how Caacupé children really used their laptops. Equally absent from the book's descriptions was evidence of the extensive work that Paraguay Educa did to promote laptop use; the presence of the trainers, the cooperation of the director, the breakdown and repairs evident on the laptops that children brought into the room, even the collection of students from across the district who were present at the school that morning went unnoted. I further realized that the didactic nature of his presentation, in addition to flying in the face of the kind of student-led exploration that constructionist learning encouraged, precluded him from even seeing what was on kids' screens. Finally, the language barrier—in addition to reinforcing a link between English and technical expertise, as we also saw in the previous chapter—added another layer of distance between him and the children. As I reflected further on this encounter, nearly everything about it puzzled me.

However, based on my notes from that day and informal conversations with Paraguay Educa employees afterward, nobody else was surprised that Bender did not see any actual classroom laptop use, even though that was how the visit had been framed. Over lunch following the classroom visit, the other members of Paraguay Educa triumphantly talked among themselves in Spanish and to Bender in English about how well the school visit went. In retrospect, the performance seemed to be so fragile, so obviously constructed—yet nobody else seemed to be bothered by this. On the contrary, nearly everybody involved seemed willing to contribute to the performance—the Paraguay Educa employees, the school director, Bender, the children themselves—and they all appeared to know what was expected of them. Bender, who seemed to be used to VIP treatment, appeared equally at ease whether expounding in front of a room of children or to a conference hall of informatics professionals. Paraguay Educa employees easily took on the role of translators and promoters. My own role as ethnographer—there to record, make sense of, and write about the events I witnessed—was meant to be one of the least obtrusive, yet when I reflected on this event later, it seemed to be the most out of place.

Reinforcing Charisma

Why would Paraguay Educa construct this visit this way, and why was Bender so uninterested in questioning it? The most obvious answer is the pragmatic one: the NGO wanted to ensure its continued survival, and a powerful figure such as Bender, if duly impressed with the project, could help with that goal. Even though Bender had left OLPC in 2008, he still headed up Sugar development and had considerable clout in this area. If the project was able to show Bender what he wanted to see, then it would be deemed successful according to its own metrics and the closely related metrics of OLPC. Of course Paraguay Educa would try to put its best foot forward. This was later confirmed in informal conversations with my host in Asunción (who had previous experience with NGOs and philanthropic projects) and other Paraguay Educa employees. The rest of Bender's itinerary, full of meeting with funders and government officials, also corroborated Paraguay Educa's hope that Bender's visit could shore up support for the project.

And the relationship that the visit cultivated with Bender did have material effects. While the Paraguayan government did not ultimately step up to support and expand the project as Paraguay Educa had hoped, Bender had already been impressed with the quality of the Sugar patches (or software fixes) coming from the NGO's talented lead programmers and gave them visibility at MIT and in the open-source-software circles he frequented. Later, he arranged for Sugar Labs to contract with several of them to continue doing software development for Sugar when Paraguay Educa's funding started to dry up. He visited more times over the years to further promote Turtle Art, and many of the pictures that were posted publicly from these events appeared similar to the event I had witnessed.

But why did the school visit in particular seem to involve obscuring much of the labor that Paraguay Educa had put into the project (as well as into the visit itself)? This is a thornier problem; in my conversations with Paraguay Educa staff and others involved with the project, nobody really articulated a reason for this. But the surprise they expressed at the question

suggests an answer. Just as technological charisma depends on an investment in the social imaginaries that undergird it—imaginaries that often go
unspoken, even as they shape thoughts and actions—perhaps an integral
part of Bender's visit was to uphold and reinforce the same imaginaries. In
particular, it needed to support the idea that the XO laptop could unlock
the innate creativity of children and harness that creativity and inclination
to play toward constructive ends, especially for the technically precocious
boys who populated the pages of Papert's writings and have been catered to
by decades of video games, over a century of technical toys, and as many
years of media portrayals. An honest acknowledgment of the incremental
results of the project and the significant work that Paraguay Educa had
done to achieve those results would undermine the power of those imaginaries, the project's charisma, and the conviction that sustained many who
were involved.

At the same time, as I read *Learning to Change the World* in 2013, I realized that Bender did know at least something about the extracurricular
clubs, trainers, technical support, and other infrastructural elements that
Paraguay Educa had put into place. The book stated, "In Paraguay, there
has been an emphasis on using the laptop during informal time; there are
many children who attend Saturday sessions that are dedicated to the creative uses of the computer."[6] Still, the passive voice obscured Paraguay Educa's work to promote these uses. Similarly, the book claimed that the parent
classes were more ubiquitous and more heavily attended than they were in
reality, and the work that went into organizing them was not even mentioned. "Parents who had rarely been seen at school flocked to afternoon
sessions about how to use the 'XO' to support their children's education,"
Bender and colleagues wrote, suggesting that the laptops flipped a switch
in previously negligent parents, resulting in them suddenly showing up at
school.[7] In another place, the book acknowledged (however vaguely) the
"superlative support" that Paraguay Educa provided—but said that it was in
support of teachers making a "leap of faith" into constructionism, neglect-
ing to discuss the fact that it was nearly impossible to use the laptops in
classrooms and that most teachers were interested not in constructionism
but in the internet:

> Stories of teachers facing difficulties making the transition from an instructionist
> to a constructionist approach to teaching populate most deployment evaluations.
> This inevitably leads to less classroom use of the laptop. For a teacher to make this

transition generally requires some training *and* a leap of faith. Since the quality and quantity of training has varied so much from deployment to deployment, the pace of change has also varied. In Paraguay, where there is superlative support, the transition is rapid.[8]

The book thus seemed to selectively understand Paraguay Educa's work as bolstering, not challenging, One Laptop per Child's underlying imaginaries: it was simply and straightforwardly enabling technically precocious children to realize their full potential, and this individual development would then lead to social and economic development in the region.

Within this charismatic frame, then, it was paradoxically in Paraguay Educa's best interest to obscure or minimize the ongoing labor and middling results that deviated from the project's vision—the breakage, the disuse, the difficult infrastructural transitions—in order to maintain and perpetuate the charisma into which it had invested so much. Given all this, perhaps what was most surprising in the vignette that opened this chapter was my own naïveté. Though they might have employed the language of assessment, Paraguay Educa and Bender were more interested in rallying people to support the project, and the two activities required completely different, perhaps even incompatible, approaches. Given that clear-eyed assessments cannot help but appear negative when lined up with a project's charismatic promises, why did I think that there would be any reason for Paraguay Educa to show Bender the true state of the project, in all its uncharismatic messiness? The complexities of the project and the significant labor that was invested to achieve even incremental results not only would have seemed paltry in the face of such utopianism, it would have threatened to undermine the very social imaginaries on which the project was built. Thus, rather than changing expectations to match reality, it *performed* an alternate reality instead.

"Perform" is, of course, a slippery word here—intentionally so. One common lay meaning comes from theater, in which a performance is a collaborative act between producers and audience, bounded by space and time, that creates an alternate reality that may speak to everyday life but is ultimately not of it. In short, it is purposefully artificial. There certainly seemed to be a level of artifice about Bender's "school tour," akin to the way that the charismatic leaders whom Max Weber studied would use performances to shore up their authority.[9] Might charismatic machines elicit performances to shore up their authority as well?

At the same time, sociologists—building off the foundational work of Erving Goffman—recognize that performances of a sort are also natural parts of our everyday social lives. Goffman borrowed theatrical terms in describing many of our social behaviors in order to capture their performative nature.[10] Moreover, "performance" and "performativity" take on broader meanings in science and technology studies, where they are understood as representations that shape social practices and meanings, through the act of internalizing and appropriating those representations. One branch of this scholarship comes out of linguistics and is focused on the performativity of language as a way in which we construct shared perceptions of reality. (Judith Butler's analysis of the ways in which gendered expressions socialize us all into particular gender roles provides an especially compelling case in this genre.)[11]

In this light, Bender's visit was performative in the ways that he and Paraguay Educa had internalized and then enacted the expectations of what a "successful" project looked like. It was a *charismatic performance*, folding together some of each of the aspects of "performance" discussed here. Its charisma made it in part theatrical, set apart from everyday experiences—evoking a charismatic potential for action more than an uncharismatic reality. However, it also expressed the understandings of the project that participants had and, in turn, reinforced those understandings. Through it all, it kept in place the social imaginaries of childhood and the model of easy cultural change on which the project's charisma was built—even if it might have compromised the longevity of the project in the process by not calling attention to its need for intensive and ongoing support.

Charismatic Performances

In the years after *Learning to Change the World* was published, as I continued to analyze and write about One Laptop per Child, I started to notice and hear about such charismatic performances everywhere in education and development projects around the world. Sometimes they were not called as much by participants but were recognized by observers. At other times, participants, too, recognized the intentionality behind the charismatic performance and the difference between it and day-to-day experiences.[12] Why did they seem to be so ubiquitous? Could this kind of charismatic performance be expected in such projects? As we have seen, they can impress funders, donors, and other project VIPs with their utopian visions, giving them the impression that projects are more successful and less complicated than they may be in reality. Like fundraiser galas, they make participants feel important for being part of such an exciting event. Moreover, the fact that Paraguay Educa could mobilize this kind of charismatic performance showed that it knew the game and could play along; it was serious about the kinds of social change that it showed in its performance and organized enough to make it happen, at least for a short time.

In fact, I realized that Bender's visit was not the only charismatic performance I had observed during my fieldwork in Paraguay. While his was the most high-profile visit, there were in fact many other visitors who elicited similar performances. When developers from Uruguay, an evaluator from Peru, and educators from Brazil visited Caacupé, Paraguay Educa put on charismatic performances for each: slideshows, roundtables, student competitions, and more performative classroom visits. One such event notable for its contrast with Bender's visit occurred just weeks after, in November 2010, at an education and technology conference in Asunción. There, in an air-conditioned hall on the outskirts of the city—a rough area known for children begging on its dusty roads—major technology companies including Microsoft, Dell, and Asus set up booths featuring flashy demonstrations of very expensive technology-rich classrooms: multitouch computer-screen tables, interactive projections, smart whiteboards, and even OLPC laptop knockoffs called Exos that featured a similarly designed handle (though they ran Windows, anathema to OLPC's commitment to open-source software). Paraguay Educa also had a booth at the event, filled with pictures of children from Caacupé smiling and holding laptops up to the camera or studiously bending over them. However, even though Paraguay Educa was the only booth representing a real project, it was one of the less charismatic amid the flashy, if completely unrealistic, performances by the larger technology players.

As an ethnographer, I was sensitive to the possibility that participants might be performing for me as well. Indeed, my presence was novel in the first weeks of my fieldwork, during which time teachers, children, and Paraguay Educa employees would sometimes show off interesting ways they used their laptops, and my presence in a room was always noted. As my fieldwork continued and I worked to make my presence routine, I found that these kinds of demonstrations largely subsided; I could just observe in a classroom without everyone's attention being on me. This also highlights the difficulty in sustaining charismatic performances: the effort needed to enact them is not something that most projects can realistically rally day to day.

On the flip side, even a visitor who was around a project long enough to see beyond charismatic performances might fall back on idealized imaginaries as memories of the uncharismatic day-to-day realities of the project fade and the imaginaries that motivated their participation take up greater importance in their lives. This happened with one of the volunteer visitors I got to know in Paraguay and with whom I kept in contact in the years after my fieldwork. During the long bus rides to and from Caacupé, I had many conversations with this volunteer about how different his expectations for the project had been from the realities he saw in Paraguay. However, in the years since, I have found that the gloss of nostalgia slowly idealized the project in his memories. Talking to him about our time in Paraguay years later, I found that he had forgotten those frank conversations and instead tended to remember the parts of his visit that confirmed the social imaginaries upon which OLPC relied. Based on my conversations with him, I do not think that this shift was an act of intentional deception on his part; it was more of an outcome of the growing physical and temporal distance from the everyday realities of the project, a distance that charisma needs. Without shifting his faith in the imaginaries that undergirded the project—a shift that would have also necessitated a frank reckoning with his own identity as a hacker and his passion for the project—it makes sense that his very memory shifted toward nostalgia instead.

One Laptop per Child itself was no stranger to charismatic performances, and the set of imaginaries that these performances upheld was fairly consistent, even as the story around the project shifted. Nicholas Negroponte, who has continued to project the image of a charismatic leader in the original Weberian sense even as memory of OLPC faded, has been especially adept at constructing charismatic performances. His first public performance for OLPC was the demonstration at the World Information Summit in Tunis in November 2005, where he and UN secretary-general Kofi Annan debuted a nonworking prototype laptop and made a number of promises regarding laptop features and project reach that OLPC was ultimately unable to fulfill.[13] Over the next several years, the charismatic performances continued

as Negroponte flung XO laptops across stages to demonstrate their ruggedness and showed pictures of smiling children sitting with their laptops in
pastoral fields and forests.[14] In 2012, as evaluations from Peru's poorly supported OLPC project showed little change, his charismatic performances
changed course: he started talking about dropping tablet computers out
of helicopters to enable children in Ethiopia to teach themselves to read
English.[15] His 2014 TED Talk evoked a vision of roaring success for this
Ethiopia project:

> I then tried an experiment, and the experiment happened in Ethiopia. And here's
> the experiment. The experiment is, can learning happen where there are no
> schools. And we dropped off tablets with no instructions and let the children fig-
> ure it out. And in a short period of time, they not only turned them on and were
> using fifty apps per child within five days, they were singing "ABC" songs within
> two weeks, but they hacked Android within six months. And so that seemed suf-
> ficiently interesting. This is perhaps the best picture I have. The kid on your right
> has sort of nominated himself as teacher. Look at the kid on the left, and so on.
> There are no adults involved in this at all.[16]

At an OLPC summit, in conversations with Negroponte's collaborator
Maryanne Wolf, and from news articles, I heard a quite different story
about this experiment.[17] The tablets had never actually been air-dropped
from helicopters with no instructors; instead, a tech-support team from
Addis Ababa had, with the help of village leaders, built a "school" with
solar panels and chargers for these devices. This team visited weekly to fix
tablets and copy the contents of their flash memory. This weekly support
crew continued to be part of the study for the first year, and adults in the
village supported the project as well. Children did exhibit some preliteracy
skills (recognizing letters) before the experiment ended, although the hacking that Negroponte referenced was finding the configuration menu and
turning on the camera. But these details hardly mattered—the charismatic
performance upheld and reinforced the imaginary of the technologically
precocious boy and was therefore often accepted by those who found that
imaginary to be important in their own lives and worldviews. Although
Negroponte's hyperboles may be easy to criticize, they are also masterful
examples of performing just what stakeholders and the technology world
more generally wanted to hear.

These charismatic performances are not unique to OLPC, of course. In
fact, the project grew out of a culture that was also built around them.

For the first twenty years of its existence, the MIT Media Lab's credo was "Demo or Die," and it was famous for such performances, as all groups housed in the lab were expected to have a demo on hand at all times for funder tours.[18] I went on one such tour myself in 2013, as part of the Digital Media and Learning summer institute. Even though the Media Lab had shifted this credo to the more concrete "Deploy or Die" several years earlier under the leadership of Joi Ito, the flashy demonstrations that graduate students readily launched into throughout the lab showed that these were still at the core of the Media Lab visitor experience. One longtime Media Lab denizen joked that even the way Negroponte would take his time flipping through his notebook to "find a slot" for high-powered donors—whether accurate or not—was also charismatically performative, evoking a schedule that was packed with even more important figures.

This culture of charismatic performance at the Media Lab made me further wonder if Seymour Papert's writings were based on charismatic performances similar to the one I witnessed during Bender's visit to Caacupé that warm October day. It is difficult to say: Papert's writing lacked discussions of research methods and indications of the reliability of his findings—the benchmarks for all good empirical research, qualitative and quantitative alike.[19] In fact, his writings derided educational research and other attempts to regularize or evaluate what he said was a highly individual and ultimately unmeasurable experience.[20] Even though scholars have critiqued the way that mass educational evaluation is commonly done, that does not necessitate an abandonment of critical analysis of alternate approaches, nor does it negate the value of accounts that are as transparent and honest as possible. Indeed, the inability of the national rollout of OLPC's precursor Logo to create a widespread computing culture in the United States in the 1980s, the failures of other constructionist one-to-one computer projects overseas (particularly Senegal in 1982 and Costa Rica in 1986), and the lackluster results of a one-to-one computer rollout in Maine that Papert had advised—which have all been held up as successful pilots for One Laptop per Child—suggests that there was some degree of charismatic performance in the ways in which OLPC leaders discussed them.[21]

Although the MIT Media Lab is especially skilled at these charismatic performances, they are also rife in the technology world more generally, from the pitches to venture capitalists to the journalists who hew close to company press releases as they cover the latest technological innovations (or speculations) with breathless enthusiasm. This is what it takes to get attention in this space; incremental benefits do not sell. On the flip side, when media outlets note something that has gone wrong with a technology, the coverage is catastrophic. It is easier to flip from utopianism to dystopianism than to wrestle with the complexities of a middle ground. But these are all charismatic performances. They reference existing cultural imaginaries in order to resonate with people and build on those imaginaries to charismatically paint a picture of a revolutionary future—whether utopian or dystopian—that the technology in question will make easy to achieve or even inevitable.

Development and education projects such as OLPC often openly admire innovation in the technology world, and they work to emulate practices from it, these charismatic performances included. But in this context, they can take on higher stakes. In a landscape of limited funding instead of generous venture capitalists, projects that perform the right imaginaries can take away resources from long-term projects that have real impacts but are not as charismatic. When performing aspirational futures has become an expectation—a yardstick against which all development projects are judged—a project that did not make such claims but was actually attainable, if incremental, can have trouble garnering support. The limited resources in this space thus can be frittered away on unrealistic and short-lived projects.

The normalization of this model in development discourse is brought home by a conference that the World Bank hosted in 2011 and 2012 called Fail Faire, which has been continued by a number of other organizers around the world under the umbrella of Fail Festivals (http://failfestival. org). The initial World Bank Fail Faires sought to give visibility to development projects that had failed—an ostensibly laudable goal, as such projects are often otherwise hushed up and hidden. However, rather than reckon with the broader patterns present in these failed development projects, these events were more focused on promoting what organizers called the "moon shot" model: projects that could be recognized from the outset as likely to fail but would be amazing if they were to succeed. In some ways, this was a marked shift, as these kinds of development projects, One Laptop per Child included, had previously tended to portray their utopian goals as easily achieved, even sure things. But it still did not question the focus on utopian goals in the first place. Although the projects featured at these Fail Faires promised unrealistically grandiose results and then inevitably failed to achieve them, the narrative was that their failures would help inform future moon shots, *not* that projects should be more realistic in scope.[22]

More recent Fail Festivals take a more measured tone in not discussing moon shots directly, instead encouraging projects to, as they put it, "fail early and often," paraphrasing a mantra that has become common in business schools and technology-management literature.[23] Still, a reliance on modeling development projects on the technology world risks taking on the reckless technological utopianism that is rife in that world as well—especially in venture capital. And the influence is one they are proud of. "Look at Silicon Valley, arguably the most successful business incubation environment in the world today," Fail Festival's website states. "Venture capitalists invest millions in new, untested ideas and are excited to get a 10% success rate. They know 90% of their investments will fail when they go to market, and that's acceptable because the 10% that succeed will be that transformative to the marketplace and their own portfolios. Now find a social development organization with that same 1 in 10 big impact mindset."[24] Projects thus show what they think the audience wants to see; they perform "transformative," they perform "1 in 10 big impact," they perform aspirational futures. This motivation to charismatically perform success stems in part from the pressure on development projects to set goals that evoke the project's charismatic promises. These promises attract the initial interest of funders, volunteers, or voters. Against the backdrop of other projects doing the same, a project that did not make such claims would have trouble garnering support.

The Catch-22 of Charismatic Performances

The ubiquity of these examples suggests that these charismatic performances are more than mere Potemkin villages—that charismatically performing development has a purpose beyond simply impressing funders with a façade of success. A more useful path for understanding the broader purpose of these kinds of performances is to consider them again within theories of performativity from science and technology studies—paying particular attention to the ways in which the representations a performance creates subsequently shape the practices of those involved in the performance. For instance, market analyses are, in the words of Donald MacKenzie, "an engine, not a camera"; they are not merely reflections of the reality of the market but actively create the market.[25] This kind of performativity is central to the ways that social imaginaries are invoked and sustained—and the ways that their subjects take up those imaginaries as their own, as integral parts of their identities. People thus may perform their understanding of a social imaginary (whether that involves seeing their own identity reflected through it or specifically excluded by it), and through this performance, the social imaginary appears to at least temporarily come true; it is reified.

With this in mind, we may ask about the vignette that opened this chapter: what reality was Bender's charismatic performance producing and for whom? When the performance ended, after all, the Caacupé children, teachers, and trainers who participated went back to their usual routines. How was the performativity in this case an "engine," and what effects did this engine have? Perhaps, in addition to being for Bender, the charismatic performance I watched was for themselves—for the employees of Paraguay Educa and the children and teachers in Caacupé. By aspirationally enacting the utopian goals of the project, these actors not only reinforced the project's foundational imaginaries but practiced for themselves what success should look like within the world they had created. Moreover, even if those involved knew on some level that the charismatic performance was hiding too much of the messy day-to-day realities of the project, perhaps demonstrating that they could mobilize to perform success showed that there was a chance that they could achieve it. In this view, Bender was performing as well: his demonstration showed what kinds of laptop uses were considered interesting, providing a model for success so that local actors could do more along the same lines. This interpretation better explains why Bender's presentation was so didactic, though he was a self-avowed proponent of constructionist learning. He was there not to witness the state of the project but to impart a vision of what the project should be.

Anthropologist William Mazzarella has described performances such as this as "beautiful balloons" that captivate—until they deflate under the weight of broken promises. And just as this book does, he has connected this captivation with a kind of charisma:

This charismatic moment, at the peak of the hype cycle, opened up new possibilities. As the moment passed, and gravity began to win, disillusionment set in. The erstwhile beautiful balloon, having apparently moved from hype to habit, remains more enigmatic than it might appear. Much of the air that has gone out of it was, in any case, stale: the exhaust fumes of nationalism, paternalist politics, and profit seeking ... And yet, even these stuffy dreams, as they unspool through new machines in old places, release potentials—potentials that may yet surprise those most heavily invested in them.[26]

In his analysis of information and communications technologies (ICT) projects in India, Mazzarella has described these kinds of performances as attempts to harness the expectations of visitors into local action toward those desires. Through "enacting" a technology and development project, the local actors he observed—like the actors in our charismatic Caacupé classroom visit performance—"were able to use the charisma of ICTs to harness the desire of external others and to convert this desire and attention into local action-potential. Although there were certainly limitations to this action potential within the field of ICT ... the charisma of ICT had the potential to 'bleed' efficacy into adjacent domains of social practice."[27]

Thus, Bender's visit to "observe" a classroom was an attempt to harness the charisma of OLPC and convert the vision that OLPC had created into (temporary) action. The performance that the visit elicited could help the project survive, even as it hid the ongoing work needed to do so; it showed teachers and students what success looked like, even as it concealed what would be required to achieve success in the long term; and it suggested that any obstacles encountered—breakage, disinterest, "little toys," drained batteries, missing software, infrastructural deficiencies, unrealistic labor expectations, design flaws, media corporations, language barriers—were merely small speedbumps on the road to laptop-driven, child-led progress. I, like Mazzarella, was fascinated by what he has called the "hype cycle," not only because of its "inflated, excessive tone" but because of the "collective desire" it demonstrated.[28] This operates simultaneously on the planes of the ideological and the material—one does not preclude the other, as the quote that opens this chapter also attests. But where, really, have these charismatic performances left Caacupé, Paraguay Educa, and One Laptop per Child? Those involved in the visit that day showed pride and pleasure in their performance, but just what "concrete social effects" have come of them?

The conversion process was—and continues to be—limited. Just as Mazzarella's participants were not given the high-paying tech jobs that were part of the promises in his field site, and just as Negroponte's Media Lab has failed to bring about its utopian visions of global community, the children in Caacupé by and large have not become the hacker-engines of economic growth for their city. Even as we met a small group of children in the previous chapter for whom contact with Asunción's elite and with international visitors was exhilarating and clearly shaped how they thought of themselves, it was less clear how this would shape their actual prospects beyond setting up utopian expectations for easy social change through laptop use—expectations that were ultimately impossible to realize, at least in the short term promised by the project. I thus came to realize that the ubiquity of these charismatic performances, coupled with the infeasibility of the charismatic visions that they work to prop up, points to something fragile in these projects. When these performances are the de facto language of development discourse, all such projects are expected to be flashy but ultimately unsustainable.

Charismatic performances can provide conviction and smooth away uncertainties, and the faith in technology they create can also foster cultural cohesion. The trouble comes when these charismatic performances become too removed from the messiness of daily life and arc of history. When we hold a charismatic technology at arm's length, rather than allowing it to cohabit the world with us and then grappling with all of its consequences, we may come to believe that a technology somehow, impossibly, transcends both historical precedent and everyday life.[29]

How does the charismatic sheen that these performances create shape what is considered sayable about these projects—the kinds of results that they can afford to make observable? When held up to charismatic promises, even changes that visiting evaluators or participants in the project would deem as positive look paltry and incremental at best, and it would be difficult to convince funders to continue to support these incremental steps when they had been promised grander transformations that would come about naturally, even spontaneously. This is the paradox of charismatic technologies. Their charisma involves, by definition, utopian promises for what cultural changes the technology will easily bring about. When even small steps toward these cultural changes are much more difficult and fraught than these charismatic visions claim—and when they need much more intensive and sustained investments to achieve—it is tempting to compare vision with reality and declare that the project had failed, even when it has incrementally positive results. Likewise, these incremental results—increased school attendance, say, or more children familiar with computers and the internet, or more children who are registered with the state—are hidden from view, discussion, and critique.

At the same time, if a project does perform success—even with all of the material and ideological effects that charismatic performances can have—this can also result in a cessation of funding. A donor might well declare that the project has achieved success—mission accomplished. This is in fact what happened with Paraguay Educa as its funding sources dried up after its first few years of intensive support with modest success. In this way, charisma is seductive, but its utopian promises become a trap, a catch-22. Charismatic performances of success reinforce the differences between what the performances temporarily create and the day-to-day realities back to which they inevitably subside. Moreover, such performances, though common in educational reform and development projects, ultimately undermine the success of these projects by not accounting for infrastructural investments, setting a realistic scope of progress, or acknowledging ongoing challenges.

So how might we approach these charismatic performances? This chapter has shown both how they are important, and why they are risky. Charismatic performances are acts of devotion, commitments to a collective ideological framework that makes social change seem easy, at least for a short time.[30] Yet participants risk getting caught in these charismatic visions, fixated on a future that may well be unattainable in lieu of a present that might be messy and difficult but is real and here. Likewise, projects in development or education that lean on charismatic performances risk compromising their long-term sustainability with them, performing the discourse of disruption in the past tense: we have disrupted, the disruption is accomplished, and we can all go home.

7 Conclusion

And so it now is up to all of us to carry forth Seymour's legacy—indeed an honorable calling ... It is the legacy of creating for children environments where each will find the gears of their childhood.

—Paulo Blikstein and Dor Abrahamson, "Logo, the Next 50 Years"

It has been well over a decade since the One Laptop per Child project was publicly announced in 2005 and more than half a century since Seymour Papert, the intellectual father of the project, began to pursue the dream of children learning with, and from, computers. Though OLPC has faded from public consciousness in that time, its legacy continues. More than two and a half million of the XO laptops that the organization designed have been distributed to children around the world, most of them in Latin America. The project is credited—whether correctly or not—with shifting popular focus from ever-faster and higher-capacity laptops to smaller, cheaper, more durable laptops or netbooks, including Google Chromebooks, which have become popular laptops to use in school computer programs.[1] On the other hand, its failures have helped shift the discourse in the development community away from the kind of charismatic rhetoric that OLPC exemplified—though these charismatic promises live on in educational technologies and across Silicon Valley more generally.

Nicholas Negroponte, OLPC's founder and public face, has started to refer to OLPC as a "valuable learning experience" in recent talks—one that did not quite turn out the way he wanted but that has nevertheless helped inform future endeavors. In 2011, Negroponte announced one such endeavor, staging a highly publicized "helicopter deployment" of off-the-shelf tablet computers to forty children in two small villages in Ethiopia, a two-hour drive from the capital city, Addis Ababa. At a conference in October 2012, he promised the same sorts of results that he had promised seven years earlier with OLPC's XO laptop: that children would teach themselves to read, that they would teach themselves and one another to hack the computer, and that they would leapfrog past the adults in their lives and become truly global, networked citizens.[2] (With these claims, we may wonder just what he had learned from OLPC.) Though the results to date have been about as lackluster as they were for projects with XO laptops, Negroponte has not seemed fazed, replacing details about the limited learning that has taken place with loaded terms such as "reading" and "hacking" that evoke a vision of much more.[3] OLPC's charisma lives on beyond Negroponte, as well. Small but dedicated groups of volunteers, many of them also advocates of free and open-source software, continue to contribute to the Sugar codebase, host local OLPC community events, and follow OLPC projects around the world.

More importantly, though, the same charismatic promises that animated this project live on, attached to other technologies in education, in development projects, and across Silicon Valley more generally. This legacy is, of course, particularly strong at the former home of Negroponte and Seymour Papert, the MIT Media Lab. Constructionism and digital utopianism—hallmarks of Papert and Negroponte, respectively—been deeply woven into the Lab's culture from its inception, as has an antischool, "I taught myself" sensibility. Papert's best-selling manifesto *Mindstorms* has long served as a core text of the Program in Media Arts and Sciences graduate course Technologies for Creative Learning (later renamed Learning Creative Learning), in which students discuss their own versions of Papert's childhood gears and use them as inspiration for their own nostalgic design. The Lab also continues to be well represented within the pages of the famously tech-celebratory magazine *Wired*, of which Negroponte was a founding investor and columnist through the 1990s.

Papert's ideas, moreover, are behind the block-based programming environment Scratch, which was developed in part by his former student Mitch Resnick and former collaborator Yasmin Kafai. This program experienced a meteoric rise alongside OLPC, especially with the debut of its popular online community for "remixing" Scratch projects in 2008. Its popularity was buoyed by being included—and, as we have seen, heavily promoted—on OLPC's XO laptops. It was moreover influenced by the same social imaginaries as OLPC and other constructionist projects—a legacy we can see in the reportedly lagging number of accounts created by women and girls in the Scratch online community, despite concerted efforts to promote it widely.[4] Scratch has been incorporated into the national school curriculum in Great Britain and has provided inspiration for the Hour of Code movement (and is the primary platform for follow-up content on Code.org), in which children across the United States are encouraged to spend at least one hour interacting with a block-based programming environment to learn some rudimentary programming concepts.

Beyond the Media Lab, the same social imaginaries that motivated OLPC and constructionism have cropped up in projects across the technology and education world. Educational reform efforts have often been techno-utopian, and charismatic technologies from radio to the internet have been hailed as saviors for an educational system that appears perpetually on the brink of failure.[5] For instance, shortly after my first round of fieldwork in 2010 concluded, the education field was abuzz with the promise—and threat—of massive open online courses, or MOOCs. Startups such as the Khan Academy, Coursera, and Udacity promised to democratize education through self-paced instruction on a computer, where students from all walks of life and from all over the world could pursue their natural curiosities even in the absence of institutional or social support.[6] Some universities embraced the move, resurrecting open courseware movements from the late 1990s with new interactive tools for customized learning, even as MOOCs were used to justify budget cuts. Others bemoaned the end of university education as we know it, pinning all of the consequences of neoliberal policies that for decades had been undercutting universities in the name of market pressures on this one new movement. But in the end, the charismatic bubble of MOOCs burst: rather than attracting diverse students and democratizing education, MOOCs have enabled highly educated professional men—a group that already enjoys ample institutional and social support—to constitute the majority of those who successfully complete these courses.[7]

These imaginaries are further accentuated within the ranks of the maker movement, which openly embraces its debt to hacker culture and hacker ideals. My work with Daniela Rosner has explored how the same social imaginary of the technically precocious and self-taught boy that animated OLPC also animates this world. This includes MIT-designed FabLabs and other makerspaces with various kinds of digital fabrication equipment;

Maker Faires where the fruits of this labor can be shared; maker-inspired private and charter schools; and *Make:* magazine, O'Reilly Media's effort to capitalize on this community. Here especially, the technically precocious boys who populated the pages of publications from a century ago, such as *The Boy Mechanic: 700 Things for Boys to Do* (1913), also populate the pages of *Make:* magazine and are targeted at maker events, where adults try to channel the same kinds of masculinized acts of rebellion into ostensibly productive ends. Even though the primary group that actually participates is older white men—members of the original hacker generation—these imaginaries of childhood nevertheless still motivate many who are involved.

There has also been a resurgence of arguments originally made in the 1970s and 1980s that all children should learn to code, and this rhetoric even featured in Barack Obama's 2016 State of the Union address. The resulting Hour of Code (for children) and coding boot camps (for adults) bring together the imaginary of the naturally creative child, the playfulness of video games, and an unschooling undercurrent that has often (though not always) situated these lessons outside of school curricula. These are taken one step further in technology-heavy charter schools, some of which are run by for-profit corporations.[8] These initiatives generally play up the same imaginaries of childhood and use parents' utopian attachments to technology and their belief in the power of programming to support (whether intentionally or not) the broader movement to privatize or dismantle public education in the United States. They also reflect the libertarian sensibilities of Silicon Valley in its rejection of public services, including some of the publicly maintained infrastructure that makes technology-centric schooling possible.

Projects more on the correcting end include flipped classrooms, virtual classrooms, and individualized student tracking enabled by video courses and grading infrastructures, such as those that the Khan Academy, Udacity, and other MOOCs have developed. Projects aimed more at dismantling or completely rebuilding, such as OLPC, include maker-inspired and project-based private and charter schools, as well as some homeschooling and unschooling movements. Like Papert and other OLPC leaders, these movements would rather scrap the existing educational system and make education a "private act" where each child finds their own way to learn or, alternatively, a "market" where the savviest students have direct access to the best ideas, without the state as an intermediary.[9] But how less privileged students might fully participate in this privatized education is often unaddressed. Although these projects—like OLPC—sometimes pay lip service to supporting diversity, they generally do not address how this individualistic proposal might enact one of the oldest promises (however unfulfilled it might be) of public education: that of school as a social leveler.[10]

These initiatives bleed into the culture of Silicon Valley more generally—a culture of "disruption" where a public infrastructure that has been defunded for decades is held in contempt by many, where formalized schooling is likewise disdained, and where stories of self-taught hackers are ubiquitous. Even though higher education is in fact the overwhelming norm among computer programmers, the narrative that circulates in six-week learn-to-code boot camps and across the industry more generally is that college and even high school are not necessary for, and might even be antithetical to, technological entrepreneurialism.[11] A small number of proudly and vocally self-taught programmers across Silicon Valley—who never finished college (or sometimes even high school) yet who developed enough technical expertise to land programming jobs or even start their own companies—help cast this exception as a norm. One, Chris Blizzard, worked for OLPC. Technology venture capitalist Peter Thiel formalized this attitude in his Thiel Fellowship. Launched in 2010, it gives one hundred thousand dollars over two years to students to drop out of college and start a company—despite mounting evidence that it is experienced middle-aged entrepreneurs who have the most success with startups.[12] More broadly, the culture of startups, venture capital, and technology development is built on visions of utopianism, as the sky-high valuations of many upstart technology companies attest.

Putting Charisma in Context

These utopian impulses are both pervasive and powerful, and the technologies that they get attached to become charismatic with these heady promises. This book contributes a framework for understanding these types of charismatic technologies, which embody and carry with them utopian visions for the kind of world that they will not only enable but make easy to achieve, even inevitable. In its original religious context, "charisma" tends to have a negative connotation: a preternatural force that hoodwinks a gullible populace. But charisma can play an important role in group cohesion by smoothing away uncertainties, contradictions, and adversities under a utopian gloss. To OLPC's contributors, the charisma of the XO laptop affirmed their belief in the power of computers in childhood, imposed coherence and direction on their work, and gave them reasons to dismiss or push back against doubters, even in the face of what might otherwise have felt like overwhelming odds or ample counterevidence.

In an interview with Nigeria's *Daily Vanguard* newspaper in 2006, Negroponte stated that "criticising this project is like criticising the church, or the Red Cross"—with the implication that anyone who would do so was a heretic.[13] We could extend this parallel to frame the fervor that many felt toward the project as a kind of faith in the charisma of the XO laptop.[14] The project's team members and contributors collectively affirmed their belief in the power of the XO to unlock children's creativity. They dismissed evidence that constructionism had previously failed to achieve its utopian goals. This imposed coherence on the project and sustained it, even as evidence mounted that its outcome might match that of previous constructionist projects. Like the charismatic religious leaders whom Max Weber studied in early twentieth-century Germany, the XO laptop exerted an uncanny holding power, connecting its acolytes to a broader purpose. It affirmed all they held dear, from the hacker ethos to the natural (technical) curiosity of children, from the irresistibility of machines to the duty of the technology industry to lead the world to a utopian digital future. Although this led to a tough reckoning among some when the laptop failed to live up to its promises, others have continued to be sustained by the same charisma attached to new causes. Indeed, the power of charisma is great enough that it can erase with the gloss of time the complications and ideological work that one may grapple with on the ground."[15]

On the other hand, charisma can also have a blinding effect. OLPC's XO laptop was charismatic to many of its boosters because it mirrored their existing ideologies—indeed, the imaginaries that were foundational to not only group memberships but their professional and personal identities—and promoted a social order with them at the top. This made it more difficult for those on the project to recognize or appreciate ideological diversity, much less constructively confront problems of socioeconomic disparity, racial and gender bias, or other issues of social justice beyond most of their personal experiences and the imaginaries that animated the project.[16] As a result, charismatic technologies such as OLPC's laptop at times work to reinforce the very inequities they are meant to alleviate.

We have seen both how OLPC's charismatic vision was built up over some fifty years of work at MIT and how it was translated and transformed when it was put into practice in Paraguay. But what is important throughout this story is not only that OLPC was wrong; equally if not more important, particularly for analyzing future projects, are the ways that its charisma remained alive.[17] It was through the ways that this charisma tapped into social imaginaries (such as the naturally creative child, the technically precocious boy, and the factory model of school) that the OLPC community understood itself and its social worlds. These imaginaries are more than mere ideology; they have very real ramifications for how funding gets allotted to education and development projects and across Silicon Valley more generally.

OLPC, like many contemporary charismatic technologies, used what Vincent Mosco calls "end of history" rhetoric: the technology is so new, so groundbreaking, that it is ushering in a radical break with the past and historical lessons no longer apply.[18] However, its charisma is not new. Even though a particular technology may not remain charismatic, its charisma may be taken up in a new technology and live on. Historians of technology have shown that over the past two hundred years, many technologies—including the railroad in the mid-nineteenth century, radio in the 1920s, and the internet today—have been hailed as groundbreaking, history shattering, and life redefining.[19] "Since the earliest days of the Industrial Revolution," Langdon Winner writes, "people have looked to the latest, most impressive technology to bring individual and collective redemption. The specific kinds of hardware linked to these fantasies have changed over the years ... [b]ut the basic conceit is always the same: new technology will bring universal wealth, enhanced freedom, revitalized politics, satisfying community, and personal fulfillment."[20]

The remarkable parallels between the hackers who designed personal computers, the internet, and OLPC's XO laptop and previous technically precocious boys are no accident. The charisma of radio, computers, the internet, locomotives, and more draws on the same set of utopian stories about technology, youth, masculinity, and rebellion. The individualist strains in these earlier technologies found voice in the cyberlibertarianism of recent decades.[21] The new frontiers of the imagination that the railroad opened in mid-nineteenth-century America were rhetorically echoed in the new frontiers of radio in the 1920s and cyberspace in the 1980s and 1990s, spaces of radical individualism and ecstatic self-actualization.[22] The charismatic appeal of tinkering that 1920s radio boys championed found voice again in computer hacking, in projects such as OLPC, and most recently in maker culture.[23] Thus, although the technologies may shift with time, the charisma lives on.

Historian Howard Segal notes that the rhetoric about the internet's power to spread democracy "was identical to what thousands of Americans and Europeans had said since the nation's founding about transportation and communications systems, from canals to railroads to clipper ships to steamboats, and from telegraphs to telephones to radios."[24] Vincent Mosco also notes these similarities in a call for both more empathy for the past and more skepticism of the present: "We look with amusement, and with some condescension, at nineteenth-century predictions that the railroad would bring peace to Europe, that steam power would eliminate the need for manual labor, and that electricity would bounce messages off the clouds, but there certainly have been more recent variations on this theme."[25] Even though it thus may be easy to discount past technologies—including OLPC—given the tarnish of time, they contain two lessons, one about the enduring importance of charismatic technologies (in whatever instantiation) to the modern cultural imagination and the other about their limits.

Taking a historical perspective on charismatic technologies can also show us both how conservative charisma is—the same kinds of promises have been made over and over, with different technologies but the same groups in power—and how unattainable its promises are. Noticing these historical parallels can help break the spell of the present. It demonstrates that today's charismatic technologies are neither natural nor inevitable and that even if they promise revolution, they repeat the charisma of past technologies and ultimately reinforce the status quo.

What could be problematic about the feelings that these charismatic technologies can evoke, at least when still relatively new? After all, many of these technologies did, in time, appear to transform the worlds in which they existed. However, Mosco argues that it was not until these technologies receded into the mundane and we understood how they could fit into the realities of daily life, rather than making us somehow transcend it, that they had the potential to become an important social and economic force.[26] Likewise, we have seen in this volume that charismatic performances of success served to accentuate the difference between the idealized world that these technologies were meant to create and the messy, negotiated realities they actually encountered, where they were in need of ongoing and intensive repair. And within the realm of education, when these messy and labor-intensive realities of learning inevitably clash with the charismatic promises that technologies make, disillusioned innovators—along with the media and the general public—often blame schools and especially teachers for blocking technological transformations that were, in reality, unattainable.[27]

Charisma, Translation, and Cultural Change

The ultimate goal of One Laptop per Child was to create cultural change. With its charismatic frame, this cultural change became a kind of cybernetic exercise, wherein laptops would automatically bring children from across the Global South into alignment with MIT's hacker culture and Media Lab worldviews. At least some in the project really did believe that all that one needed to do was equip children with laptops and that the children would then take care of cultural change themselves, apart from or even in spite of the social institutions and cultures in which they were embedded. Negroponte explained in a 2010 editorial in the *Boston Review* that "owning a connected laptop would help eliminate poverty through education. ... In OLPC's view, children are not just objects of teaching, but agents of change."[28] Statements such as this sound as though they give agency to children, but within this is also a sobering individualist responsibility: if change fails to materialize, it is not the fault of the schools or economic conditions or social structures or national policies or infrastructure—those have already been written off. It is the fault of the children.

This model of technology-driven cultural change was fairly common in other education and development projects—and, at one point, in the social sciences more generally. In the 1960s, communication theorist Marshall McLuhan supported the idea of technological determinism, arguing that cultural change comes about because of key technologies such as writing, printing, and electronic media.[29] Scholars of science and technology would, of course, challenge this story by interrogating the cultural contexts and negotiations that enable the development of these technologies and take place around each of them.[30] Yet McLuhan's sentiment lives on elsewhere. Following the promise of technologically induced change, modernist development projects from the 1960s onward plunked technologies and engineering projects down in impoverished regions with expectations of easy cultural transformation instead of plans for support or maintenance.[31]

More recently, one high-profile initiative, Sugata Mitra's Hole in the Wall project, has claimed that children can teach themselves a whole host of things via computers installed into literal holes in the sides of buildings in villages across India, with no instruction whatsoever. Although accounts of this have not been independently verified and, in fact, a number of researchers doing fieldwork in India have reported seeing these Hole in the Wall computers revert to just holes in walls, Mitra and the Hole in the Wall computers are often held up alongside Negroponte and OLPC as proof that children can lean on their natural curiosity and do not need formal instruction to learn. (Solidifying the connection, Negroponte wrote the foreword for Mitra's e-book on self-directed learning.)[32] Across these examples, the story went like this: give out the device, maybe provide a little bit of support, but let the device speak for itself. Users will surely see the benefits it could provide them.

OLPC's and constructionism's faith in children to direct their own learning seems admirable, leaning on social imaginaries of children as nobler, purer, and more natural than the meddling adults in their lives. Indeed, part of a charismatic technology's allure is the promise of easy cultural change. A charismatic technology, like a charismatic human leader, makes change not only compelling but seemingly effortless, natural, even inevitable. By the same token, opposition becomes unnatural or even immoral, as evinced by Negroponte's claim that criticizing OLPC was akin to criticizing the church or the Red Cross. Ironically, though, charismatic allure is based on appealing to an underlying ideological status quo and elevating this ideological frame—in OLPC's case, the worldviews of the technology industry and particularly its social imaginaries of childhood, school, and technology—as the way to a utopian future.

The social sciences have spent a century and a half studying the processes of cultural change, and their findings challenge this vision. "Culture," in these worlds, remains a contested term linked to collective ideologies and imaginaries, social groups and institutions, and particular social positions and experiences. In anthropologist Clifford Geertz's words, as part of various cultures, subcultures, and countercultures, we are "suspended in webs of significance that [we ourselves have] spun."[33] And shifting these webs is no small task.

Moreover, in light of OLPC's mission to bring these ideas to children across the Global South, constructionism and OLPC could also be seen as imperialist, and Paraguay Educa's faithful adherence to OLPC's vision as problematic.[34] After all, Paraguay Educa uncritically adopted a set of ideals largely developed at MIT, an elite institution in a country with a history of both military and cultural imperialism in the region: the United States. It moreover chose to invest in an untested technological intervention instead of food, vaccinations, working bathrooms, or any number of other kinds of much-needed aid. OLPC's claim that a country could merely replace its textbook budget with a laptop budget and access online textbooks, for instance, assumed that the target country had a textbook budget to begin with and that schools in that country regularly used textbooks. In Paraguay, for one, this was not true.

Thus, OLPC's popularity in Latin America sits less comfortably with the history of other technology projects in the area. Early in the project, a number of scholars critiqued OLPC for its cultural imperialism, highlighting the way that project leaders themselves spoke of exporting their own ideals and values through the laptop—even as these leaders sometimes claimed it was also "value neutral" because of its open-source software.[35] Indeed, many Latin American countries are rightfully suspicious of the cultural and military power that the United States holds in the region: it has supported coups and dictators, and its corporations have polluted their landscapes, starved their children, and tried to charge whole countries for their own drinking water.[36] The 2014 volume *Beyond Imported Magic* (which contains a chapter on OLPC in Paraguay) and work by other Latin American scholars in science and technology studies have complicated the imperialist notion that technology simply flows from the Global North to passive and grateful recipients in the Global South.[37] At the same time, they have also demonstrated how technological utopianism can obscure problematic elements of development projects, such as corporate greed, inequitable distribution, neoliberal policies, and wasted resources.[38]

In Paraguay, we found that one of OLPC's main avenues for enacting cultural change was through the imagination. The charismatic power that the XO possessed came not so much from the realities of the laptop in use as from the kinds of desires that the laptop as symbol was able to fabricate. For Paraguay, a country shaped by significant humanitarian intervention, it is remarkable that this project could even partially command this charisma, even when other local projects—schools sponsored by American churches, Peace Corps projects, religious mission work, and more—did not seem to. The XO laptop inspired hope for (internet-enabled) change by signifying various possible futures to the teachers, children, and especially Paraguay Educa employees most invested in the project.

Thus, perhaps one of the enduring legacies of OLPC is its symbolic power to connect to dreams for a technology-rich future. That these dreams persisted in spite of the object's less glamorous material realities reveals the moral claims made about new information technologies more generally, as well as the role that charismatic objects play in maintaining such contradictions. The lasting impact of the XO laptop may be its ability to reveal the wide gulf between the ideals and realities of joining the global information society that played out not only in the use of the laptop but in visions of how it might shape the region's or country's future—from the constructionist learning that Paraguay Educa promoted to the portal to the English-centric internet that some children and teachers embraced.

However, there are consequences to charisma. The pressure that educational or social reformers are under to produce charismatic projects that can attract attention and funding also leaves them in a situation stacked toward failure, a system that rewards showy but myopic projects. This research suggests another course by documenting the tremendous amount of work—social, infrastructural, and ideological—needed to produce even incremental social change. As NGOs like Paraguay Educa take on more of the functions formerly expected of states (such as registering students, enforcing vaccinations, and providing schooling), moderating expectations, promoting transparency, and learning from failures become ever more important.[39]

Paraguay Educa pursued the dream of an OLPC project to promote quality education, economic growth, and civic engagement in Caacupé. In light of Paraguay's uncertain political future, declarations of technology's role in creating good citizens took on an even more idealistic tone, in much the same ways that the rhetoric around Peru's open-source-software movement produced a vision of its role in civic responsibility.[40] But the ways that the organization and those using the laptops discussed the future of the project tended to embrace the utopian rhetoric common in OLPC and elsewhere that children would naturally "take charge of their own learning" and embrace futures as globalized information workers.[41] Against this backdrop, transnational corporations were moving into OLPC projects in Latin America to take advantage of a new avenue to access this market of young protoconsumers.

In these visions, the child learner becomes the perfect neoliberal subject, given full agency in their education and, by extension, full responsibility. Whereas the mythology often invoked here is one of freedom, the flip side of responsibility becomes much more salient in use, as laptops break down with little recourse for repair, as media-centric use proves much more interesting than more creative endeavors, and as advertising from transnational corporations edge in on children's attention and loyalties in the name of education. In the process, they delegitimize the role of educational institutions and rarely acknowledge that learning is a fundamentally social process.

These trends risk both naturalizing and worsening inequalities that projects like OLPC are meant to eliminate—particularly gendered, socioeconomic, linguistic, and center-periphery divides. This might in fact be the path of least resistance: if OLPC's laptops settle into the mundane under these forces, they are likely to do so primarily as media consumption devices, much as radio and television (both also once the stuff of utopian daydreams) have done in the United States.

Harnessing Charisma without Getting Ensnared

One of the goals of this book has been to expose the ideological stakes that buttress charismatic technologies. Those who create, study, or work with technology ignore the origins of charisma at their own peril—at the risk of always being blinded by the next best thing, with little concept of the larger cultural context that technology operates within and little hope for long-term change. Recognizing and critically examining charisma can help us to understand the effects it can have and then, if we choose, to counter them.

To that end, one of the aims of this book, and of this concluding chapter in particular, is to "make the familiar strange," to echo a common mantra in interpretive research.[42] Identifying and analyzing a technology's charisma helps us recognize the ideological commitments that technology makes and perhaps even our own ideological commitments, which may otherwise be as invisible as water is to the proverbial fish.[43] And if one can identify the ideological commitments of the technology world, then that is a first step in avoiding getting caught up in them.

The case of One Laptop per Child shows us why it is dangerous to ignore the origins of charisma: one risks being perpetually entranced by the newest charismatic technology. This is not to say that cultural change with a technology-centric project is impossible. Still, even more realistic reforms grounded in the realities of their intended beneficiaries sometimes have difficulty gaining broad popular support outside the school unless they add a charismatic gloss of rapid, revolutionary change.

This charismatic pressure can put even open-eyed reformers in a catch-22.[44] They must promise dramatic results to gain the social and financial support for reforms, and then they must either admit to not achieving their goals or pretend that they did achieve them. Either way, funders will declare that the project is finished and withdraw financial support, and then researchers and other observers will begin to note the discrepancies between reformers' promises and their own observations. Thus, projects that rely on charismatic technologies are often short lived; their resources are cut off before charisma recedes into the background and before the technology becomes part of everyday classroom experience. This catch-22 has dogged efforts for educational reform, development, and cultural change—especially those funded through grants or other short-term funding—for well over a century. As the technology community moves on to the next charismatic device without learning from its failures, this will continue to hamper the possibility of real, if incremental, change.

Moreover, as long as we are enthralled by charisma, we might actually prevent these technologies from becoming part of the messy reality of our lives as we wait for them to help us transcend it. After all, charisma is ultimately a conservative social force. Even when charismatic technologies promise to quickly and painlessly transform our lives for the better, they appeal precisely because they echo existing stereotypes, confirm the value of existing power relations, and reinforce existing ideologies. Meanwhile, they may divert attention and resources from more complicated, expensive, or politically charged reforms that do not promise a quick fix and are thus less charismatic.

What is the alternative to this trap of charismatic, technology-driven cultural change? In the space of educational reform, David Tyack and Larry Cuban argue that "tinkering," or incremental and locally situated reform, is more effective in the long term, even if it is not charismatic. "It may be fashionable to decry such change as piecemeal and inadequate, but over long periods of time such revision of practice, adapted to local contexts, can substantially improve schools," they explain. "Tinkering is one way of preserving what is valuable and reworking what is not."[45] This is a far cry from the technology world's rhetoric of disruption, but it has a better chance of being effective.

Likewise, an antidote to charismatic thinking lies in Donna Haraway's call to "stay with the trouble," which is what this account has tried to do throughout. We can acknowledge the importance of charisma, social imaginaries, and technological utopianism but also recognize that utopianism is at best a dialectic, not an aspiration. It is most useful as a method for interrogating, understanding, dwelling in, and possibly improving upon the present, not an excuse to ignore it (much less the past). "In the face of such touching silliness about technofixes (or techno-apocalypses)," Haraway reminds us, "sometimes it is hard to remember that it remains important to embrace situated technical projects and their people. They are not the enemy; they can do many important things for staying with the trouble."[46]

The intention here is not to advocate for an eradication of ideologies; just as it is impossible to escape the bounds of our own bodies and subjective points of view, so too is it impossible to operate entirely outside ideological frameworks. However, a large body of Marxist theory notes that becoming aware of the ideological frameworks in which we operate allows us to evaluate whether they are really serving the purposes we hope or assume they are.[47] Only by way of this cognizance can we shift them if they are not.

The conclusions of this line of inquiry help us understand the use of charismatic imaginaries in technological design and the effects, whether intended or no, that this charisma-centric design may have. This research has traced the multifaceted charisma of the XO laptop across its design and its use, but we have also seen that the XO is not the only technology to be built around charismatic ideals. Indeed, many technologies have, or at least start with, an element of charismatic authority in them. Only by understanding the origins and effects of this charisma can we hope to recognize it, and only by recognizing it can be hope to harness it, open eyed, for good.

Appendix A   An Assessment of Paraguay Educa's OLPC Project

In collaboration with Paraguay Educa, the nongovernmental organization (NGO) in charge of the project in Paraguay, I designed two exams that tested reading comprehension and mathematical reasoning. We administered these two exams to all third- and sixth-grade students in the program in 2010 (who received laptops in April 2009, in Phase I of the project), a group of students about to join the program (who would receive laptops in May 2011, in Phase II of the project), and a control group in nearby schools. We administered the same tests in the same schools in 2013. There was one large difference in the program between 2010 and 2013. In 2010, Paraguay Educa's project was in full swing in all Phase I schools, with full-time teacher trainers in every school and a full-time technical-support staff. In 2013, the program had been expanded into Phase II schools, but the teacher-trainer program had been discontinued and the support staff significantly cut back due to lack of funding.

Methods

In collaboration with Paraguay Educa's director of education, María de la Paz "Pacita" Peña, we developed and validated two multiple-choice exams of thirty questions each—one for third-grade students and one for sixth-grade students—that tested basic literacy (i.e., identifying the main character and understanding the storyline of a short story) and basic numeracy (i.e., relationships between numbers, spatial relationships, and basic sequences). In November 2010, we administered these tests to all third- and sixth-grade students in the program, all third- and sixth-grade students about to join the program, and a comparable and similarly sized group of third- and sixth-grade students from schools nearby that had no plans of joining the program, for a total of 2,085 students. We again administered the same tests to the same schools in November 2013, with 2,565 students. By testing all third- and sixth-graders in the experimental conditions and an approximately equal number in a control group in 2010 and 2013, we have more than enough data points for statistical significance and to eliminate some of the biases that can occur when studying only a subset of students.

Phase I and Phase II schools differed widely in their educational attainment before the laptop program: schools in Phase I were generally larger, more urban, and wealthier than Phase II schools. The control group had a mix of school types similar to Phases I and II combined. Thus, even when trying to control for public versus private, large versus small, and urban versus rural schools, comparisons between Phase I and Phase II schools likely include confounding factors, though we did conduct comparisons between each phase and the control group. More reliable is the within-groups comparison that we were able to make between 2010 scores and 2013 scores for the three conditions: Phase I, which had laptops in both 2010 and 2013; Phase II, which did not yet have laptops in 2010 but had them in 2013; and the control, which did not have laptops for either test. Below we list a number of specific hypotheses that we test with the data.

We selected third and sixth grades as suitable benchmarks in the Paraguayan schooling system for several reasons. By the end of third grade, all students are expected to be able to read and write basic Spanish, and at the end of sixth, they are promoted to junior high and out of the official laptop program (though students continue to own their laptops and Paraguay Educa has administered some opt-in extracurricular programs for secondary-school students).

The tests were multiple choice, with four choices per question. Roughly one-third of the questions were about reading comprehension—which consisted of one fiction passage and one nonfiction passage—and two-thirds were about mathematical reasoning (equal parts numeracy, patterns/algorithms, and spatial reasoning). Questions were modeled after cognitive reasoning tests for students two grade levels lower in the United States and after the 2006 Sistema Nacional de Evaluación del Proceso Educativo (SNEPE), a standardized test previously used in Paraguay and elsewhere in the world. We chose to model on tests two grade levels lower because that is the level at which Paraguayan students have historically performed on standardized tests such as the SNEPE compared to students in the United States, and we did not want our results to bottom out. Like the high-school-level Programme for International Student Assessment (PISA) test, questions were developed specifically to not depend on any specific curriculum (e.g., words or concepts taught in a particular grade) but instead were designed to test general cognitive reasoning. They were also written to be locally appropriate; for example, one short reading passage discussed Paraguayan food.

We validated the third-grade test once and the sixth-grade test twice with the help of an Asunción-based education-evaluation expert, Dr. Oscar Serafin. We administered both tests to pilot groups of third-grade and sixth-grade students in October 2010, and Dr. Serafin conducted a validation of the responses. At the end of the validation, each test had a Cronbach's alpha score of over 0.7, which is the minimum threshold for reliability of the test. The school that piloted the tests in 2010 was excluded from the analysis as well as the counts below.

In 2010, Paraguay Educa's teacher trainers and other employees administered the test on paper to all third- and sixth-grade students, totaling 2,085 students in fifty-two schools. The testing took place November 15–19, the end of the 2010 school year in Paraguay, just before summer vacation started. There were three groups of students who took the test: 576 students in Phase I schools who had received laptops in 2009 (treatment 1), 773 students in Phase II schools who would receive laptops the following May (treatment 2), and 724 students in the nearby towns of Atyrá and San Bernardino, which had no plans for joining the laptop program (control). The control group included similar proportions of public and private, large and small, and urban and rural schools as the experimental groups. In addition to providing a between-groups comparison of results to date, this 2010 test provided a baseline measure for a follow-up test with third- and sixth-grade students at the same schools at the end of 2013, five years into laptop use for Phase I schools and three years in for Phase II schools.

In 2013, Paraguay Educa's teacher trainers and other employees—a subset of the same group that administered the test in 2010—administered the same test on paper to all third- and sixth-grade students in the same fifty-two schools, this time totaling 2,565 students. The 2013 testing spanned the two weeks of November 4–8 and 11–15, near the end of the 2013 school year in Paraguay. This time, 848 students were in Phase I of the program (treatment 1), 916 students were in Phase II of the program (treatment 2), and 801 students were in schools in nearby towns not in the program (control).[1] Paraguay Educa's employees and I manually entered and cross-validated the 2,085 returned tests in 2010 and the 2,565 returned tests in 2013. We excluded blank tests for consistency, resulting in 2,071 of the 2,085 tests given in 2010 and 2,472 of the 2,565 tests given in 2013 being included in the analysis.[2]

I conducted Welch's two-sample t tests to evaluate whether the mean exam scores were statistically significant between various groups (between 2010 and 2013 within-groups and between treatment and control groups).[3] These results were broken down by subject (language or mathematics), by grade (third, sixth, or combined), by the year the test was administered (2010 or 2013), and by other factors that could influence scores, including whether the school was public or private, urban or rural. I used the statistical analysis package R to conduct the analysis and to generate density plot graphs to visually observe the difference in means and variances.

Results for Phase I Schools

Phase I schools had laptops in both 2010 and 2013. In 2010, they also enjoyed the support of teacher trainers and technical-support staff, support that had mostly been discontinued by 2013. Doing a within-groups comparison of Phase I scores in 2010 and 2013, the only moderately significant ($p < 0.05$) difference was that sixth-grade math improved by 3.3% in 2013 as compared to 2010. Because Phase I schools were enrolled in the laptop program in both 2010 and 2013, these results did not clearly tell us what the effects of the laptops were. However, they did tell us that if Paraguay Educa had made a difference to this group in 2010, it appeared to continue to do so in 2013. On the other hand, in the control group, where we expected no difference between 2010 and 2013, we did in fact observe one moderately significant ($p < 0.05$) difference between 2010 and 2013: a decrease of 3.9% in sixth-grade mathematics.

A between-groups analysis of the 2010 and 2013 tests, which compared Phase I students with the control group, showed modest differences between Phase I students with laptops ($N = 529$) as compared to the control group—though these results could not suggest causality. Looking at just the 2010 test and comparing Phase I and control students, we found a moderately significant ($p < 0.05$) positive difference in third-grade math, with Phase I students scoring 3.5% higher than the control. However, we also found a moderately significant ($p < 0.05$) negative difference in sixth-grade math, with Phase I students scoring 4.3% worse than the control group in 2010. We did not find any statistically significant differences in third- or sixth-grade reading scores between Phase I and the control in 2010. Despite their moderate statistical significance, these results suggest that there were no strong differences between Phase I and control students in aggregate in 2010.

For the 2013 test, between Phase I and the control we found a strongly significant ($p < 0.01$) positive difference in scores in third-grade math, with Phase I schools scoring 5.3% better than the control. We also found a moderately significant ($p < 0.05$) positive difference in scores in third-grade reading, with Phase I schools scoring 3.8% higher than the control. Other subjects and grades did not have statistically significant differences in the 2013 test.

Overall, these findings suggested that there were incremental benefits to being part of Paraguay Educa's OLPC laptop program, especially for mathematics: the within-groups analysis showed gains in sixth-grade mathematics in 2013 as compared to 2010, and the between-groups analysis showed differences in third-grade mathematics in both 2010 and 2013. However, we acknowledged the imperfection of studying Phase I schools, which had laptops for both tests, and the many issues of between-groups comparisons in particular. We attempted to reduce such issues by including in the study a similar number of students from urban and rural schools and large and small schools, but there could be other factors for which we did not account. Furthermore, we restricted this analysis to public schools, as public schools were funded and provisioned similarly across the sample, whereas private schools varied widely in wealth and quality.

Moreover, it was impossible to know what actually caused this difference. Perhaps it was tinkering with the laptop itself, but perhaps the teacher training and the lectures and suggestions of visitors from OLPC and elsewhere provided teachers with excuses to reflect on pedagogical practices and to be encouraged to try new classroom arrangements and more child-centered learning models, as well as encouraging ongoing conversations around what worked best, which teachers were having the most success, and why. Even if some of these changes were due to a placebo effect motivated by the laptop's presence, the results of the Phase I tests in 2010 and 2013 still suggest that this program made a small positive difference.

Results for Phase II Schools

The test results in Phase II schools were meant to provide the most solid evidence of the effects of the laptop. For the test in 2010, the twenty-five Phase II schools did not yet have laptops, and these scores established a baseline. We did find that the mean 2010 scores in Phase II schools were about 6% lower than the mean scores in the control group in all subjects and grade levels tested (third-grade reading, third-grade math, sixth-grade reading, and sixth-grade math), with all of those differences statistically significant ($p < 0.01$). We hypothesized that this substantial lag was due to the relatively high number of rural schools in Phase II as compared to the mix of schools in the control.

By the time of the 2013 test, students in Phase II schools had had laptops for two and a half years, including most of the 2011 school year, all of the 2012 school year, and all of the 2013 school year. However, there were significant changes in Paraguay Educa during this time as well. Phase II schools had one year of part-time teacher-trainer help in 2011 and then two years with little or no systemic support from Paraguay Educa. We found that in 2013, the gap between Phase II schools and the control schools had narrowed, with mean scores between 1% and 4% lower than the mean scores in the control group for all categories, but the results were not consistently significant.

By conducting a within-groups comparison of the 2010 and 2013 scores in the Phase II schools, we were able to more directly establish how reading and mathematics scores changed with the introduction of the laptops. However, we did not see many changes. The only moderately significant ($p < 0.05$) difference between 2010 and 2013 Phase II test scores was that third-grade reading improved by 3.1% in 2013 as compared to 2010. We did not observe a difference in reading in sixth grade or mathematics for either grade. This was in contrast to Phase I schools, where we saw a consistent difference in third-grade mathematics in both 2010 and 2013. We hypothesize that Phase II schools did not benefit more from the program because much of the social support structure, which needed ongoing financial support, had been discontinued after the first year of the Phase II program.

Discussion

The strongest results were the within-groups comparisons between 2010 and 2013. Some of these scores improved slightly, though the only moderately statistically significant difference was in third-grade reading, which improved 3.1%. We found that the relatively higher scores in the Phase I schools and the control schools mostly held steady between 2010 and 2013, with the exception of sixth-grade mathematics, which had a moderately statistically significant ($p < 0.05$) improvement of 3.3% in Phase I schools and decrease of 3.9% in the control schools. It is also worth noting, however, how much lower the baseline (2010) scores of the twenty-five Phase II schools, which were predominantly small and rural schools, was: around 6% lower across all grades and subjects than the ten relatively larger and more urban Phase I schools and the control schools. This again illustrates the structural differences between rural and urban students in Paraguay.

Based on the lack of more significant change between 2010 and 2013, we hypothesize that the initiatives that were discontinued, though expensive, were very important to the success of the program as a whole. Although the 2013 results were disappointing, they did indicate that the expectations that often govern education and development projects continue to emphasize short-term gains over long-term sustainability, despite ample evidence that such approaches do not work. These findings have implications for scholars, policy makers, and (perhaps most importantly) funding agencies who establish the scope of education and development projects.

Appendix B Methods for Studying the Charisma Machine

A comprehensive account often calls for a multiplicity of research methods and data sources. Although my overall approach is interpretivist, and my methods predominantly ethnographic, I have incorporated texts, statistics, surveys, and controlled experiments over more than a decade of researching One Laptop per Child. My sources for the first two chapters include speeches about OLPC; discussions on public mailing lists, wikis, and discussion boards; interviews with some developers; and publications about constructionism, the educational theory behind OLPC. To explore OLPC's development process, I analyzed OLPC's publicly accessible online records, most of which were created between 2006 and 2009. This included its wiki (http://wiki.laptop.org), its public mailing lists (http://lists.laptop.org), the personal blogs and web pages of its employees, publicly accessible talks that employees gave, and news articles about the organization. I found that the motivations that OLPC contributors had for joining the project and their visions for its future have been captured in considerable detail in these places, and from them I have been able to construct a detailed picture of the motivations behind and logistics involved in the development process. The people involved include around twenty full-time OLPC employees, who were most central to shaping OLPC's direction, and hundreds of open-source contributors, a few of whom were particularly influential. In October of 2010, 2011, and 2012, I attended the annual One Laptop per Child Community Summit in San Francisco, a conference of OLPC employees and volunteers, to update these findings.

Emails to OLPC's mailing lists were automatically compiled and posted by the mailing list program Mailman to http://lists.laptop.org. All but a few of the mailing-list archives were public, with no account or sign-in needed to view them. Though OLPC maintained over seventy public mailing lists, just a few lists accounted for most of the list traffic in 2007 and 2008. Topical lists, such as "Games" and "eToys," provided information about specific aspects of OLPC's content, and a number of lists such as "OLPC-Peru" discussed OLPC projects in specific countries.

I also conducted six interviews with OLPC contributors in 2008 and several between 2010 and 2016, but I have found that these provided information that has also been expressed online and have used little of their content here. Most respondents seemed to be very aware of OLPC's public face and stuck with the party line when talking with me, even when they were more frank on OLPC's publicly archived mailing lists. My timing also made interviews difficult: by 2008, when my research began, many core employees had quit and did not respond to queries or were not interested in being interviewed. Thus, I have drawn more heavily on analysis of online content instead of interviews.

Chapters 3 through 6 draw on data I collected during seven months of fieldwork with an OLPC project in Paraguay: about six months from mid-June to mid-December 2010 and an additional month of follow-up research in November 2013. During fieldwork, I observed children, their teachers, and their families in classrooms, their homes, and public spaces, typically spending two to three days per week in schools. Much of this time was spent observing classrooms in the ten schools with laptops. In 2010, this included four thousand students and teachers with laptops, which they received in April 2009, in ten of the thirty-six primary schools in the area. Students in the remaining twenty-six primary schools received laptops in May 2011, and teachers in those schools received them along with 150 hours of training in July 2010, during my fieldwork. My fieldwork began with shadowing the teacher trainers hired by Paraguay Educa to provide full-time curricular and technical support in the schools. I also shadowed Paraguay Educa's technical-support staff, which during my time in Paraguay alternated between visiting the ten schools with laptops to do repairs, devoting one visit per week to each, and installing wireless routers and school servers in the twenty-six schools that had not yet received laptops. As my fieldwork progressed, I spent ample time visiting classrooms without Paraguay Educa staff present to establish myself as an observer independent of that organization.

Although I spent an average of three days per week throughout my time in Paraguay in the schools and community with the XO laptops, this average was not evenly distributed throughout my fieldwork. I observed classrooms more often in my early fieldwork, between late July (when classes started up again after winter break) and mid-September (when I left to visit Peru for a week), spending pretty much every day that school was in session in the schools. When I returned from Peru in late September, I visited schools two days per week through October but spent other days collecting and analyzing quantitative data and helping Paraguay Educa develop the cognitive test described in Appendix A. In late October, I briefly visited the United States (for the OLPC Community Summit) and Uruguay. In November, I returned to Paraguay and visited schools about three times per week until the school year ended in late November. Then I briefly visited Uruguay for a second time, for the Ciudadanía Digital (Digital Citizenship) conference, before returning to Paraguay again. My fieldwork continued several weeks after the school year ended until mid-December, during which time I helped digitize the results of the exams we gave, observed a summer camp put on by Paraguay Educa, conducted more interviews, and visited the school district office.

In addition to visiting schools while class was in session, I attended Paraguay Educa's teacher training sessions; visited various community meetings in Caacupé; attended school festivals and other special events, as well as promotional events sponsored by Paraguay Educa; attended leadership workshops for the teachers most innovatively using the XO in the classroom; spent time in the parks and public spaces of Caacupé; and visited residents in their homes with the help of an excellent research assistant who was very familiar with the town and its residents.

In 2010, this amounted to about forty-nine full school days, or approximately four hundred hours, observing classrooms and schoolyards. I added another eight school days of observations in 2013. I took notes and, when in public spaces or given permission, photographs or short videos. I scaled my number of observations to the number of teachers and students at the school. thus, I observed more often at the largest schools, rotating between different areas at each of these schools, and less often at the smaller schools. Through this, I aimed to observe all classrooms and school areas as equally as possible. Depending on the proximity of schools and from whom I could catch rides, I would either spend all day at one school or visit one school for the morning session and a different school for the afternoon session. Each school-session observation, both morning and afternoon, included before-school play (which I occasionally had to miss in the morning if I was coming from Asunción and my bus arrived late), three and a half hours of classes, half an hour of recess, and after-school play.

When not in Caacupé, I spent time in the Paraguay Educa office in Asunción and was hosted by a Paraguay Educa employee not far from this office. In addition to organizing and processing my field notes, I attended meetings, observed office culture, and helped with various tasks. I also socialized after hours with Paraguay Educa staff, particularly with my host, who rented a room in her house to me.

As part of my fieldwork in Paraguay, I conducted 144 interviews with children, parents, teachers, administrators, Paraguay Educa employees, and others involved in the project. In 2010, I interviewed 123 people, including 109 residents of Caacupé (40 children and 69 adults, including parents, teachers, school principals, administrators, and other community members) and 14 Paraguay Educa employees. (Brief informal chats with children or teachers in the schoolyard are not included in this interview count and were far more numerous. I do not have recordings or transcripts of these brief conversations but do have field notes that describe them.) In 2013, I interviewed another 18 children, 1 parent, and 2 administrators. Most of the 109 interviews in Caacupé were with people associated with the ten schools in Phase I of the project, who had laptops in 2010. I also interviewed some teachers, administrators, and families in Phase II schools, who did not yet have laptops in 2010 but did know about the program and had usually encountered laptops, either in training sessions or via relatives or friends. I interviewed a proportional number of teachers from all schools, as well as all school directors who agreed to be interviewed. My teacher selection included at least one teacher who was enthusiastic about the laptop program at the school (and usually all such teachers whom I could identify), at least one teacher that was skeptical, and at least one teacher who was undecided or neutral. (There was at least one of each of these at all ten schools in Phase I that had laptops during my fieldwork, even at the small schools.) I would identify these teachers through my observations and by talking to the teacher trainers who were hired by Paraguay Educa, all of whom I interviewed in my first couple weeks of fieldwork.

In each school with laptops, I used several sources to identify students who were particularly proficient or creative with the laptop. I discovered most from my school observations and from either interviewing or informally chatting with teachers, directors, teacher trainers, and other students. In all interviews with teachers, teacher trainers, and directors, I asked about which students were most skilled with the laptops. My research assistant was invaluable in helping me track down these students and schedule interviews with them. Most of these students agreed to an on-the-record interview, although some did not (generally due to shyness). I was, however, able to observe all of these especially proficient students in the classroom and schoolyard. In total, I was able to interview about twelve of them and their caretakers (and often their siblings as well).

Based on the school's student population, I would also ask a number of other students in each school—drawn roughly at random from the student population—if they were willing to be interviewed. I aimed to have equal representation between genders and grades and proportional representation of each school, which I roughly achieved. This group of interviewees included some children who used the laptop for media purposes and some who were not interested in using the laptop much at all outside the classroom. The results of these interviews corroborated and provided explanation for my schoolyard observations of students who used laptops and students who did not.

Interviews ranged from ten minutes, for particularly shy kids or some teachers on break between classes, to ninety minutes, with an average length of thirty minutes. With the consent of both the children and their parents or guardians, I audiotaped all interviews and, if given additional permission, videotaped one or two minutes of children's use of the laptops. I preferred to conduct interviews in participants' homes—both for their convenience and for contextual information about their lives in particular and life in Paraguay in general—though a small number of participants opted to be interviewed at school instead.

Although I did have a list of questions as a guide and memory aid, these interviews were all open ended and conversational. I followed leads that participants provided rather than sticking to a fixed order of questions, which made the interactions more natural and, in many cases, ended up answering most of the questions on my list in the process. I used the questions more as a checklist to ensure that I had covered all of the key topics that I wanted to cover in each interview. I developed these questions based on both my initial research interests and my first few weeks of fieldwork observations in Paraguay.

For 73 of the 123 interviews I conducted in 2010, I had the help of a wonderful research assistant who grew up in Caacupé, knew many in the town, and was fluent in both Spanish (which I also spoke) and Guaraní (which I did not). Even though many participants spoke Spanish well, there were some rural families whose children spoke primarily Guaraní and were very shy with me, and this assistant proved indispensable. She was also a wonderful source of local knowledge throughout my fieldwork. She and another Caacupé-based research assistant transcribed all of my interviews, after I provided training and software, and translated anything in Guaraní into Spanish for me, which I then translated into English. Nearly all interviews were in Spanish, with the exception of a few with Paraguay Educa employees who spoke fluent English and a few rural students who spoke mostly Guaraní. I transcribed the interviews in English myself. My assistant's husband provided rides to many of the more remote houses and schools.

I supplemented my ethnography with some quantitative data, which I used to triangulate results in my analysis. In August 2010, Paraguay Educa gave me access to the full record of laptop breakage reports and repairs filed by its support staff. These breakage statistics were all recorded in an Inventario system custom built and released on SourceForge by a Paraguay Educa developer. Paraguay Educa kept detailed accounts of the current owner and current state of each laptop, updated by their technical support staff, who rotated between the schools on a weekly basis asking to see laptops that needed repairs. This data included types of breakage for each laptop in the program, allowing me to layer on additional levels of analysis, including the laptop owner's gender (categorized by their name, before I anonymized the data set), grade (provided in the database), and whether they lived in the urban center or the rural outskirts of Caacupé (determined by which school they attended). Thus, my reports on breakage numbers are not estimates; they are actual statistics for the entire project, drawn directly from the full set of reported breakages in Paraguay Educa's database in August 2010. The technical support staff were motivated to accurately report breakage and repair numbers in order to justify their own existence to Paraguay Educa. However, as laptops could be monitored remotely when they checked in with the school server to obtain a new security certificate, the staff could not overreport working laptops as unusably broken. Children were also motivated to report breakages in order to get their machines repaired, which the repair team could generally accomplish on the spot for software problems and the few hardware problems that did not require replacement parts (e.g., loose connections). My own time shadowing the repair team corroborates their accurate reporting. These reports were also used to order repair parts from Uruguay when possible.

In September 2010, I collaborated with the Paraguay Educa software development team to summarize laptop-usage statistics for all laptops that had been backed up on the school servers through an aggregate analysis of the Journal entries. The Journal program was meant to be a record of all programs opened and the last state of all projects, listed in reverse chronological order—much like the Recently Opened lists that some computer programs or web browsers keep—but for every program in the Sugar environment. In theory, all laptops should have been backed up to the school servers. In practice, however, some laptops (and one entire school server at a small school) were misconfigured and did not create backups, some laptop Journal programs were manually cleared by their owners, and quite a few of the programs that children installed themselves were not recorded in Journal. Thus, these records were not a perfect reflection of actual use, and I did not rely heavily on them in my analysis. The code that we used to scrape the laptop-usage logs from the school server backups is freely available on GitHub, with an explanation of the code on the Paraguay Educa wiki (http://wiki.paraguayeduca.org/index.php/Analisis_de_Uso_de_Actividades). Despite the fact that we were not able to progress as far as we had hoped with this analysis, in part due to my primary collaborator leaving Paraguay Educa for a job in Silicon Valley, we still generated school-level counts of all programs used.

From the school district, I collected classroom-level attendance data from 2008 (before laptops were given out) and 2009 (the year they were distributed) for every classroom in the ten primary schools that had laptops in 2010. I surveyed approximately 350 Caacupé teachers about their use of other technology and their experiences with OLPC's laptop. Nearly all Phase I teachers (150) and Phase II teachers (200) participated. Finally, I collaborated with Paraguay Educa's director of education in developing and administering two multiple-choice tests of thirty questions each, one for third-grade students and one for sixth-grade students, that tested reading comprehension and mathematical reasoning. The results of these tests—which we ran with all third-grade and sixth-grade students in the program and a nearby control population in November 2010 and again in November 2013—provided a controlled assessment of student learning in reading and mathematics. These results are presented in Appendix A.

To contextualize my findings from Paraguay, I also spent two weeks in Uruguay and one week in Peru in 2010 to familiarize myself with those OLPC projects. I have also followed reports and news articles about other projects from afar. In Uruguay, I visited three elementary schools with XO laptops; interviewed five people involved with Uruguay's government-run program, called Plan Ceibal; and attended the Ciudadanía Digital conference hosted by Plan Ceibal, where researchers presented findings about the deployment there. In Peru, I interviewed the government official in charge of the OLPC project for the current administration and his assistant, a contractor evaluating the program for the Inter-American Development Bank, a professor who was critical of OLPC in his country, and an open-source software programmer who was an early contributor to Peru's OLPC project. In both Uruguay and Peru, I found it difficult to get beyond the party line with officials or to be granted permission to conduct independent research. Interested spectators of the projects in both countries suggested that a close examination of their programs by outside researchers was not in the best interests of the national governments, particularly of the reelection campaigns of the officials involved with the program.

All data were obtained in compliance with Stanford University's Institutional Review Board, and the entire manuscript underwent a thorough review by a fact checker prior to publication.

...

What did fieldwork look like? I often began my schoolyard and classroom observations by drawing a map in my notebook of where the people I was observing were and what they were doing, as well as the location of furniture and other materials (desks, blackboard contents, etc.). I would pay particular attention to the locations of laptops: ones that were visible but closed, ones that were out and open but unattended, ones that were used by one child, and ones that were being used or watched by more than one child (noting the dynamics between the children). I noted what was on their screens. I would circulate among children, jotting down more details about what they were doing, sometimes briefly chatting with them, and noting new developments—the arrival of an ice-cream seller in the school-yard, for example, or the commencement of a lesson.

One of the goals of fieldwork and a distinguishing feature of an ethnography is what anthropologist Clifford Geertz calls *thick description.*[1] A thick description goes beyond appearances to an explanation of what those appearances signify to those involved; as Geertz says, it distinguishes the significance of a wink from a blink. In my fieldwork, I strove for careful observation and a balance between enough familiarity with the culture under study to recognize the significance of various behaviors (to know the meaning of the wink) and enough distance to notice behaviors that may seem natural or otherwise unremarkable to those involved (to take note of the wink as something to record in the first place).

One weakness of thick description is that anything that is subject to interpretation is also subject to *mis*interpretation; as with everything, one's assumptions and expectations color what one sees and what those things signify. Although this risk is present in any scientific study, qualitative or quantitative, I always strove to identify and test my interpretations against other alternatives, explicitly asked people about my and their assumptions when possible, validated across multiple observations and with many people, and corroborated with other sources of data to create a more complete, and rigorous, picture.

It was a running joke in my research assistant's family that we were the "laptop missionaries," because we walked around town as a pair, knocking on doors and talking to people about their laptop use. However, this mis-characterized my commitment to telling the story of laptop use as I saw it, both the good and the bad, without being viewed as either an evangelist for OLPC or an employee of Paraguay Educa. In fact, I actively worked to create distance between myself and Paraguay Educa in the minds of Caacupé residents so they would be more honest with me about their opinions of the laptop. Though officially a volunteer with the NGO, I worked independently and spent much more time in Caacupé than most other Paraguay Educa staff, except for the locally based teacher trainers and repair staff. With trusted access to schools and guarantees of participant anonymity, I was able to see and hear many local opinions of the project that the Asunción-based NGO could not. I made sure to explain in every interview that I wanted participants' honest opinions, good or bad. I also made sure to ask about what problems had come up with the laptops and, in later interviews, asked about specific problems that had emerged as themes in early interviews, such as concerns about distraction in the classroom, breakage, and control of internet content. At the same time, I could not change my identity as an outsider coming from an elite institution in the United States. Though I made every effort to ensure otherwise, I acknowledge that it is fully possible that the participants I interviewed still focused on what they thought I would want to hear. For this reason, I have corroborated my findings from interviews with my field observations and other data whenever possible. Throughout my work, I have also striven to be reflexively honest about my embodied positionality and the potential consequences of that on my results and among those I have studied.

Notes

Introduction

1. The term "Global South" refers to those countries and regions, largely but not entirely in the Southern Hemisphere, that have been under the thumb of colonialism in the past and might still be subject to extractive neocolonial regimes. "Global South" is a less normative alternative to "Third World" (which has its roots in the Cold War's anticommunism) and "developing countries" (which often assumes that "development" is a linear process and "developed" a desired outcome).

2. Brand, Media Lab; Hassan, "MIT Media Lab."

3. Rheingold, "Slice of Life."

4. See Sunstein, *Republic.com*; Sunstein, *Republic.com 2.0*; Sunstein, *#Republic*; Turner, *From Counterculture to Cyberculture*.

5. Markoff, "Taking the Pulse of Technology."

6. *MIT News*, "Annan Presents Prototype."

7. On his blog, Media Lab professor Ethan Zuckerman later reported that ten minutes of cranking yielded thirty minutes of use. See Zuckerman, "Child's Play"; Poulsen, "Negroponte."

8. Twist, "Debut for $100 Laptop."

9. Kirkpatrick, "First Working $100 Laptop"; Twist, "Debut for $100 Laptop." Also see Poulsen, "Negroponte" (who quotes Negroponte as saying, "It's every child in the world whether they want one or not. They may not know they want one"); Shreeve, "Hand-Cranked Computers" ("As Negroponte told the Summit, his aim is to put an internet- and wi-fi-enabled laptop in the hands of every child in the developing world by 2010").

10. Shreeve, "Hand-Cranked Computers."

11. Negroponte, "Hundred Dollar Man"; Bullis, "Hundred-Dollar Laptop"; Poulsen, "Negroponte"; *MIT News*, "Annan Presents Prototype"; Knight, "$100-Laptop Created"; Getty Images, "UN World Summit on the Information Society Photos: Negroponte."

12. See, e.g., Shreeve, "Hand-Cranked Computers"; Smith, "$100 Laptop"; Farivar, "That $100 Laptop." Mike Ananny analyzed ninety news articles about the laptop published in 2005—the project's debut year—and found that this press largely bought into the technology-centered, free-market-driven development agenda of the project, even with a near complete lack of details about how children would learn and a sometimes-hostile characterization of the Global South (Ananny, "$100 Laptop"). Brendan Luyt also analyzed Negroponte's marathon of talks through 2005 and 2006, along with the larger social forces shaping the early meanings of the project, including countries' professed desires to train a technologically literate workforce through the project and technology companies' desire to access new markets. Luyt found that all of these meanings bought into the techno-romantic belief in technology as salvation, a belief that this analysis recognizes as well (Luyt, "Negotiation of Technological Meaning").

13. Ananny and Winters, "Designing for Development." Several African countries, including Ghana, are intimately familiar with these kinds of concerns. See, e.g., Burrell, *Invisible Users*; Winston, "Let Them Eat Laptops."

14. *UN News*, "UN Supports Project"; Negroponte, "Negroponte at WEF."

15. Negroponte, "One Laptop per Child"; Negroponte, "Negroponte at WCIT"; Negroponte, "Hundred Dollar Laptop"; Negroponte, "Negroponte Keynote"; Negroponte, "OAS - Presentation"; Negroponte and Walsh, "Negroponte at NECC 2006"; Stahl, "What if Every Child"; Negroponte, "OLPC Analyst Meeting"; Negroponte, "Negroponte and the $100 Laptop"; Negroponte, "Negroponte at World Bank."

16. Half of this two million dollars went to the project and half to the Media Lab. Bullis, "Hundred-Dollar Laptop"; Bender et al., *Learning to Change the World*, 34.

17. Bender et al., *Learning to Change the World*, 44.

18. This was a standard part of most of his 2006 talks. See, e.g., Negroponte, "One Laptop per Child."

19. Vota, "Only Hope to Eliminate Poverty." Though Negroponte presented an easy target by often saying outrageous things that sometimes left even others on the project shaking their heads or backpedaling, many on the project echoed variations on this particular theme.

20. Mosco, *Digital Sublime*, 30–31.

21. Nye, *American Technological Sublime*, 23; Mosco, *Digital Sublime*, 22–24.

22. Dourish and Bell, *Divining a Digital Future*, 3.

23. Anderson, *Imagined Communities*, 6.

24. Taylor, *Modern Social Imaginaries*, 23.

25. Anthropologist Arjun Appadurai, who theorizes culture and globalization in the context of contemporary India, explains that "imagination has become an organized field of social practices, a form of work (both in the sense of labor and of culturally organized practice) and a form of negotiation between sites of agency ('individuals') and globally defined fields of possibility" (*Modernity at Large*, 31).

26. Jasanoff and Kim, *Dreamscapes of Modernity*, 4.

27. Coontz, *Way We Never Were*; Mansell, *Imagining the Internet*, chapter 7.

28. Appadurai, *Modernity at Large*, 53–54.

29. Berlant, *Cruel Optimism*.

30. Marx's take on ideology, for instance, provides a framework for understanding the difference in power between the capitalist elite and the workers who enable their riches. See Marx and Engels, *German Ideology*.

31. Hall, "Rediscovery of 'Ideology,'" 61, 67.

32. Dobrin, *Constructing Knowledges*, 123.

33. Hall, "Rediscovery of 'Ideology,'" 72.

34. Weber, "Charismatic Authority."

35. Several scholars have mentioned charismatic objects or charismatic technologies, though most use "charisma" colloquially, without detailing just what the theory means in their particular context. For example, Leopoldina Fortunati describes technologies as charismatic in her work, though she does not specifically discuss what it means for objects to have charisma (see, e.g., Fortunati, "Mobile Phone"). Similarly, William Mazzarella—whose work helps anchor the penultimate chapter of this book—describes technology-centered development projects as charismatic but, like Fortunati, leans on colloquial meanings of "charisma" without specifying more precisely what he means by the term (Mazzarella, "Beautiful Balloon"). Maria Stavrinaki, a scholar of art history and theory, has described the Bauhaus "African Chair" as a "charismatic object" within the Bauhaus community, an object that perfectly embodied the movement's aspirations to blend aesthetics and functionalism into a vision for a modernist future for post–World War I Germany and beyond (Stavrinaki, "African Chair"). Jessica O'Reilly engages more deeply with charisma in discussing how scientific data, especially climate data, are often presented as charismatic in public advocacy debates (O'Reilly, *Technocratic Antarctic*). Likewise, anthropologist Anna L. Tsing also engages more deeply with retheorizing charisma to show how

the idea of globalization became charismatic in the 1990s, just as the idea of modernity had in the 1950s. Academics are also taken in by this charisma, Tsing argues, and uncritically naturalize and reinforce globalist agendas by characterizing globalization as universal and inevitable. Tsing's model of charisma—a destabilizing force that both elicits excitement and produces material effects in the world (even if these effects differ from those that were promised)—is at play here as well (Tsing, "Global Situation"). My own work on "charismatic technology" was first published in 2014 and further elaborated in 2015 (see Ames, "Translating Magic"; Ames, "Charismatic Technology").

36. Using a concept from STS, we might also interpret OLPC's laptop as a kind of *boundary object* that serves as a point of interaction between various groups: its designers; the broader technology community following (and supporting) the project; the governments and nongovernmental organizations buying, distributing, and maintaining the laptops; and the teachers, parents, and especially the children who were meant to be its beneficiaries. OLPC's laptops are indeed subject to a high degree of interpretive flexibility, the most cited aspect of boundary objects; they meant various things to each of these groups. There are other aspects of boundary objects, however, that fit OLPC less well. Boundary objects are useful in detailing the organizational practices, breakdowns, and repair work of work groups. In the words of scholar Susan Leigh Star, who coined the term, "My initial framing of the concept was motivated by a desire to analyze the nature of cooperative work in the absence of consensus. ... The dynamic involved in this explanation is core to the notion of boundary objects" ("Not a Boundary Object," 604). For OLPC, however, each of the four groups I listed above largely had consensus internally and rarely worked with other groups, at least not directly. Thus, unlike the amateur and professional ornithologists interacting over the same specimens in the same lab space that Star analyzed, there was no "boundary work" for a laptop as boundary object to mediate. To connect the project and its laptop to theories of social worlds and social change, I thus look elsewhere.

37. Winner, "Do Artifacts Have Politics?" As sociologist John Law noted in 1991, "In the last decade, ... sociology has started to take both the body and the text extremely seriously ... [but] machines have been excluded from most of the new enthusiasms. It is not that there *is* no technology in sociology. Indeed, there is a real sense in which sociology *assumes* the presence, the active operation of (say) the technical. ... [But] technology does not appear to be productively integrated into large parts of the sociological imagination. Since Foucault, we have no difficulty in inscribing texts on bodies, or constituting agents discursively. But (with a few notable exceptions) it does not occur to us to treat machines with the same analytical machinery as people" ("Introduction," 7–8). Moreover, Law argues, "most sociologists treat machines (if they see them at all) as second class citizens. They have few rights. They are not allowed to speak. And their actions are derivative, dependent on the operations of human beings" ("Introduction," 16).

38. See, e.g., Johnson, "Mixing Humans and Nonhumans."

39. Jasanoff, Kim, and Sperling, "Sociotechnical Imaginaries," 3. In this way, we are all hybrids: every technology is developed and used within a myriad of social contexts, and technological constraints and possibilities likewise shape our social worlds. Also see Latour, *Reassembling the Social*.

40. See Akrich, "De-scription of Technical Objects"; Law, "Introduction"; Jasanoff, Kim, and Sperling, "Sociotechnical Imaginaries."

41. Gay et al., *Doing Cultural Studies*.

42. Might "fetish" fill the same role as "charisma"? Certainly fetishism has been invoked by scholars studying the near religious fervor attached to some technologies, such as Apple's iPhone, or "Jesus Phone" (see, e.g., Campbell and La Pastina, "iPhone Became Divine"). However, analysis via "fetish" tends to focus on an object's physical form, rendering it passive and emphasizing its materiality as a source of power. In a discussion of the "fetish of technology," David Harvey asserts that "the fetish arises because we endow technologies—mere things—with powers they do not have" ("Fetish of Technology," 3). This materialistic use of "fetish" in the social sciences has a long history: Karl Marx's "commodity fetish" invoked the religious connotations of fetishism to critique the transformation of the social process of production into the monetary "value" of a product (see Hornborg, "Technology as Fetish").

43. McIntosh, "Weber and Freud," 902.

44. This potential for charisma to override ostensibly rational thought is something that is not lost on marketers. Although they may not call it as such, their promotions often tap into the charisma that certain technologies have, and journalists often echo it as well. In fact, religion scholar William Stahl has examined the popular discourses about technology and concludes that "our language about technology is implicitly religious" (*God and the Chip*, 3). Even though charismatic technologies hence possess and reinforce their own charisma, the media is often implicated in amplifying it (see Ames, Rosner, and Erickson, "Worship, Faith, and Evangelism").

45. Watters, "Click Here to Save Education."

46. See Ceruzzi, "Moore's Law and Technological Determinism"; Dafoe, "On Technological Determinism"; Humphreys, "Technological Determinism."

47. Kirsch, "The Incredible Shrinking World," 542.

48. See Douglas, *Listening In*; Nye, *American Technological Sublime*; Mosco, *Digital Sublime*; Winner, "Technology Today."

49. For more, see Winner, "Do Artifacts Have Politics?"; Winner, "Opening the Black Box"; Bijker, "Sociohistorical Technology Studies."

50. McIntosh, "Weber and Freud," 902.

51. Mosco, *Digital Sublime*, 15.

52. Wilson, Pilgrim, and Tashjian, The Machine Age in America, 355, as quoted in Kling, *Computerization and Controversy*, 52.

53. See, e.g., Dunne and Raby, *Speculative Everything*.

54. See, e.g., Dourish and Bell, *Divining a Digital Future*.

55. Mosco, *Digital Sublime*.

56. Tyack and Cuban, *Tinkering toward Utopia*.

57. Toyama, *Geek Heresy*; Sims, *Disruptive Fixation*.

58. Haraway, *Staying with the Trouble*, 1.

59. See Derndorfer, "OLPC in Paraguay"; Bender et al., *Learning to Change the World*, 6.

60. In 2010, I interviewed 123 people, including 109 residents of Caacupé (40 children and 69 adults including parents, teachers, school principals, administrators, and other community members) and 14 Paraguay Educa employees. In 2013, I interviewed another 18 children, 1 parent, and 2 administrators. Most interviews were in Spanish, which I speak with moderate fluency; a few were in Paraguay's second official language, Guaraní, with the help of my research assistant; and a few were in English. See Appendix B for more detail on methods.

61. See Vota, "'Trojan Horse' Comparison"; Ananny, "$100 Laptop."

62. Law, "Introduction," 12.

Chapter 1: OLPC's Charismatic Roots

1. Berstein, "Low-Cost Laptop Program."

2. At a presentation for the World Bank with Walter Bender, Negroponte elaborated on how decades of working with Papert had inspired his work on the project: "The two of us took our experience, in my case, over thirty years of it working with one particular person named Seymour Papert, whom some of you may know of ... in an approach to learning which is generally called constructionist" (Negroponte and Bender, "New $100 Computer").

3. Negroponte, "Hundred Dollar Laptop."

4. Papert, "Digital Development."

5. Schofield, "Seymour Papert Injured."

6. Negroponte was previously a student at MIT, where he obtained his bachelor's and master's degrees in architecture.

7. Papert, *Mindstorms*, 19, 23, 208.

8. Quoted in Echikson, "Microcomputer Center in Paris." In this same article, another source wondered whether Negroponte and Papert might have joined together because of President Reagan's deep cuts to research funding in the United States.

9. Sullivan, "Man behind Logo," 70.

10. OLPC, "Vision: History."

11. Negroponte, "$100 Laptop"; Negroponte, "One Laptop per Child"; Papert et al., *Logo Philosophy and Implementation*, ix–x, 2–21; Bender et al., *Learning to Change the World*, 22–23, 175–180; Papert, The Children's Machine, 75–78; Warschauer, *Laptops and Literacy*. Additional overseas Logo projects by students and colleagues of Papert are described in Papert et al., *Logo Philosophy and Implementation*.

12. OLPC Wiki, "OLPC Myths."

13. On Edinburgh University, see O'Shea, "Mindstorm 2"; Koschmann, "Logo-as-Latin Redux"; Boulay and O'Shea, "How to Work the Logo Machine"; O'Shea and Self, *Learning and Teaching*; Self, *Microcomputers in Education*. On Kent State University, see Clements, "Logo in Education"; Clements and Sarama, "Research on Logo"; Yelland, "Mindstorms"; Clements and Gullo, "Effects of Computer Programming." On the Bank Street College of Education, see Pea, Kurland, and Hawkins, "Development of Thinking Skills"; Kurland and Pea, "Children's Mental Models"; Pea, "Cognitive Technologies"; Pea and Kurland, *Logo Programming*; Pea and Kurland, "Cognitive Effects"; Pea, "Symbol Systems."

14. In fact, some researchers found that prior abilities in mathematics or spatial reasoning predicted facility with computer programming—whether with Logo or other languages—rather than programming improving these abilities, pointing the arrow of causality the other way. See Webb, Ender, and Lewis, "Problem-Solving Strategies"; Yelland, "Mindstorms." For a more detailed survey of these results and their implications, see Ames, "Hackers, Computers, and Cooperation."

15. "By the end of the Center's first year, Papert had quit, so had ... Nicholas Negroponte," a 1984 *Datamation* magazine article reported. The remnants of the project were "a battlefield, scarred by clashes of management style, personality, and political conviction. It never really recovered. The new French government has done the Center a favor in closing it down." Tate, "Blossoming of European AI," as quoted in Lawler and Yazdani, *Artificial Intelligence and Education*, 7.

16. Dray and Menosky, "New World Order," 15.

17. Thornburg, "Whatever Happened to Logo?," 83.

18. Papert, "Computer Criticism," 28.

19. Papert, "Computer Criticism," 24–26.

20. Papert, *Mindstorms*, viii.

21. Papert, "Children's Machine," 21, 83.

22. Lego, "History – Mindstorms"; Walter-Herrmann and Büching, *FabLab*; Resnick et al., "Scratch: Programming for All"; Resnick, "Next Generation of Scratch."

23. See, e.g., Resnick, *Lifelong Kindergarten*.

24. Negroponte, *Being Digital*, 204.

25. See, e.g., Guernsey, *Minds of Babes*; Martinez-Gomez et al., "Sedentary Behavior and Blood Pressure"; Page et al., "Children's Screen Viewing"; Dunckley, "Gray Matters"; Pappas, "Exercise Doesn't Make Up"; Bilton, "Jobs Was a Low-Tech Parent."

26. Papert, *Mindstorms*, viii, 9–11, 18.

27. Papert, *Mindstorms*, 11, 23, 32, 182; Papert, *Children's Machine*, 67, 166, 179, 188. At times Papert instead used "tool-to-think-with," e.g. Papert, *Mindstorms*, 27, 131, 172; Papert, *Children's Machine*, 161.

28. Papert, *Mindstorms*, 26–27; Papert, *Children's Machine*, 13; Papert, "Computer Criticism," 28.

29. Papert, *Mindstorms*, 16.

30. Papert, *Children's Machine*, 14, 33–34, 64, 158, 171. Papert further writes, "Early designers of experiments in progressive education lacked the *tools* that would allow them to create new methods in a reliable and systematic fashion. With very limited means at their disposal, they were forced to rely too heavily on the specific talents of individual teachers or a specific match with a particular social context. ... When educators tried to craft an actual school based on these general principles, it was as if Leonardo had tried to make an airplane out of oak and power it with a mule" (*Children's Machine*, 14–15).

31. The "hacker" has been the subject of considerable scholarship, which generally conceptualizes the term in one of three ways: as a symbol of a particular value system or ideology, as an identity intertwined with new forms of work, or as a contested public persona. In *Hackers*, Steven Levy's approach is to define the hacker as symbol (Levy, *Hackers*). More recently, cultural scholar Douglas Thomas has used Levy's work along with other primary documents and accounts, such as "The Hacker Manifesto," to further explore the implications of a hacker ethos (see Thomas, *Hacker Culture*; The Mentor, "Conscience of a Hacker"). This ethos does not exist in a vacuum; other scholars have explored the interplay between these values and the economic, social, and legal conditions in which they exist. Gabriella

Coleman examines this context in detail, charting the heterogeneous movements that fall under the umbrella of "hackers" and showing that Levy's account is specific to MIT rather than being indicative of hackers more broadly—though it is at least legible across much of the technology world (see, e.g., Coleman, "Hacker Conference"; Coleman and Golub, "Hacker Practice"; Coleman, *Coding Freedom*). Tying this hacker ethos to the material world, cultural historian Fred Turner shows how the hacker identity was actively constructed at the first Hackers Conference in 1984 in relation to Levy's "hacker ethic," the 1960s counterculture, and certain forms of work (Turner, "Digital Technology"). Bringing in the social and legal ramifications of hacking, critical media theorist Helen Nissenbaum discusses how this hacker identity is contested and has changed from being positive and countercultural in the 1980s to one associated with criminality outside of hacker circles in the 1990s due to changes in the computing industry, hacker values, laws, and popular culture (Nissenbaum, "Hackers and the Contested Ontology of Cyberspace"). Taking a different route, Swedish philosopher Pekka Himanen redefines the "hacker ethic" as a play-based work ethic in contrast to the Protestant work ethic (Himanen, *Hacker Ethic*). The ethos that originated at MIT, which is central in this account, has influenced groups far beyond those hackers. It was independently developed (with some variation), Levy says, on the West Coast among a group of "hardware hackers" and would go on to be shared by many across the computer hardware and software businesses in Silicon Valley. It drove the industry's famous utopianism, as demonstrated by Turner and Thomas, and became intertwined with its culture for years to come.

32. Papert, *Mindstorms*, viii.

33. Papert, *Mindstorms*, 7.

34. Papert, *Children's Machine*, 33.

35. Papert, *Children's Machine*, 13.

36. Papert, *Mindstormss*, 22–23, 112–115.

37. See Kidwell, "Elusive Computer Bug."

38. Papert and Harel, "Situating Constructionism."

39. See Kessen, "American Child"; Mintz, *Huck's Raft*.

40. See, e.g., Lazar, Karlan, and Salter, *101 Most Influential People*, 133–135, 176–178.

41. Amy Ogata argues that this "primitivist" view of children's creativity is in particular a modernist conceit, which placed "modern" civilizations at the peak of the path to development and sophistication. See Ogata, *Designing the Creative Child*.

42. Philosopher John Locke is often credited with starting this shift from child-as-original-sinner to child-as-tabula-rasa. His treatise *Some Thoughts Concerning Education* encourages parents to enable a "healthy" childhood in order to make a better

adult. Jean-Jacques Rousseau similarly described the benefits of freedom in child-hood in *Émile; or, Concerning Education*.

43. See Chudacoff, *Children at Play*; Mintz, *Huck's Raft*; Zornado, *Inventing the Child*.

44. Middle-class white American parents in the post-WWII era became preoccupied with what they understood as their children's creative needs: more personal space, more opportunities for play, and toys that promised to boost IQs and provide oppor-tunities for the creative exploration of a child's passions that might translate to future careers. "Inverting the norm of training children to assume the conventions of adulthood," Amy Ogata argues, the focus on creativity during this postwar era "valued the child's unique insight, which the parent labored to reveal, sustain, and then emulate" (*Designing the Creative Child*, xi). Also see Bernstein, *Racial Innocence*.

45. Papert, *Mindstormss*, 29, 51. See also: "Papert's vision was that children should be programming the computer *rather than being programmed by it*." Blikstein, "Sey-mour Papert's Legacy," emphasis in original.

46. Papert, *Mindstorms*, 26.

47. The imaginary of the "digital native" walks the line between portraying tech-nology as a positive or a negative influence in children's lives, but it does borrow tropes from the imaginary of the naturally creative child in framing technology use as something that children can be native to. See Palfrey and Gasser, *Born Digital*; Selwyn, "Digital Native"; Ames, "Managing Mobile Multitasking." For OLPC-related examples, see Negroponte, *Being Digital*; Papert, *Mindstorms*; Papert, *Children's Machine*.

48. See Dupuy, *Origins of Cognitive Science*; Edwards, "Closed World"; Turner, *Demo-cratic Surround*; Turner, *From Counterculture to Cyberculture*.

49. Papert, *Mindstorms*, 169–171.

50. See Dupuy, *Origins of Cognitive Science* for more on the history of artificial intelligence.

51. Papert, *Mindstorms*, 61, 114; Papert, *Children's Machine.*, 35–56, 137–156.

52. Papert, *Mindstorms*, 50.

53. Papert, *Mindstorms*, 8.

54. Papert, *Children's Machine*, 24.

55. Papert, *Mindstorms*, 133.

56. Papert, *Mindstorms.*, 132.

57. Papert, *Children's Machine*, 2–3.

58. See, e.g., Papert, *Mindstorms*, 7.

59. See, e.g., Papert, *Mindstorms*, 19, 30.

60. Papert, *Mindstorms*, 31–32.

61. Papert, *Mindstorms*, 115.

62. See, e.g., Negroponte, "One Laptop per Child."

63. Negroponte, "No Lap Un-topped."

64. Bender, "One Laptop per Child."

65. Bender et al., *Learning to Change the World*, x.

66. Negroponte, *Being Digital, 200–205*; Negroponte, "One-Room Rural Schools."

67. Projects more on the reconstruction end include flipped classrooms, virtual classrooms, and individualized student tracking enabled by video courses and grading infrastructures, such as the ones that the Khan Academy, Udacity, and other massive open online courses (MOOCs) have developed. Projects aimed more at dismantling, beyond OLPC, include maker-inspired and project-based private and charter schools, as well as some homeschooling and unschooling movements.

68. Marxist educational theorist Paulo Freire, whom Papert and others in OLPC often invoked as justification for their anti-school rhetoric (though they did not address the differences between his radical "praxis"—combining theory and practice— and their technology-enabled anti-school sentiments), instead called this the "banking model of education," where students are treated like empty vessels to be filled with knowledge. See Freire, *Pedagogy of the Oppressed, 57–60*; Blikstein, "Travels in Troy with Freire," 23. This theme is explored further in chapter 3.

69. Though a crude measure, the Google Ngram Viewer shows only two books in its index that mention the "factory model of education" before 1990 but hundreds that mention it between 1990 and 2017. Although this in part reflects the corpus that has been indexed, it also highlights the anxiety behind the commoditized shift in education after the publication of the federal report *A Nation at Risk: The Imperative for Education Reform* in 1983, as described in Tyack and Cuban, *Tinkering toward Utopia*, 1–2, 115, 144.

70. Papert writes, "Thus, not only do good education ideas sit on the shelves, but the process of invention is itself stymied. This inhibition of invention in turn influences the selection of people who get involved in education. Very few with the imagination, creativity, and drive to make great new inventions enter the field. Most of those who do are soon driven out in frustration. Conservatism in the world of education has become a self-perpetuating *social* phenomenon" (*Mindstorms*, 37).

71. Papert, *Mindstorms*, 8–9.

72. Papert and Freire, "Future of School."

73. For instance, as a counterpoint, Paul Willis's classic educational ethnography *Learning to Labor* demonstrates that it was the performative rejection of school among working-class "lads" in the United Kingdom—though a performance that was in many ways expected and even co-performed by their teachers—that set them up for the culture of working-class factory jobs. In contrast, those who did well in school ended up as their managers.

74. See, e.g., Tyack and Cuban, *Tinkering toward Utopia*; Nasaw, *Schooled to Order*; Rury, *Education and Social Change*; Cuban, *Oversold and Underused*; Cuban, *Teachers and Machines*; Cuban, *Blackboard and the Bottom Line*; Cuban, "How to Break Free."

75. Tyack and Cuban, *Tinkering toward Utopia*, 111.

76. Levy, *Hackers*, 29, 31.

77. Levy, *Hackers*, 31.

78. Papert, Mindstorms, vii, 7, 9, 19, 40, 50, 96, 114–115, 174; Papert, *Children's Machine*, ix, 1–5, 13, 140; Papert et al., *Logo Philosophy and Implementation*, ix.

79. Negroponte, *Being Digital*, 3, 225–226; Blizzard, "OLPC Analyst Meeting"; Hill, "Geek Shall Inherit the Earth."

80. See Mosco, *Digital Sublime*, 2, 80, 91–93, 124, 130–131; Oldenziel, "Boys and Their Toys"; Mintz, *Huck's Raft*, 82–83, 221, 302, 317–319; Chudacoff, *Children at Play*, 38, 39–66.

81. The roots of this rebellion, like the roots of the technically precocious boy and the naturally creative child, lie in the Romantic-era prescriptions of what constitutes good play. In response to adults' attempts to control and direct play, Howard Chudacoff shows, children have always developed strategies of agency and resistance that allow them to use play for their own ends, ideally outside of adults' purview or at least without their explicit direction (*Children at Play*, 39–66).

82. Lienhard, *Inventing Modern*, 192.

83. Turner, *From Counterculture to Cyberculture*; Ensmenger, "Signs of Rugged Individualism."

84. Badham, *WarGames*; Lisberger, *Tron*.

85. See Winner, "Technology Today"; Turner, "Digital Technology"; Turner, *From Counterculture to Cyberculture*; Mansell, *Imagining the Internet*.

86. See Kling, *Computerization and Controversy*; Nye, *American Technological Sublime*; Segal, *Technological Utopianism*; Winner, "Technology Today"; Mosco, *Digital Sublime*; Turner, "Digital Technology"; Turner, *From Counterculture to Cyberculture*.

87. Hill, "Geek Shall Inherit the Earth."

88. Papert, "Digital Development."

89. Papert, *Mindstorms*, 37.

90. Amy Ogata writes, "The inventiveness of the savage boy inventor, the presumed reader of handbooks and popular science manuals, derived from both manual competence and a streak of productive experimentalism. At once virile, and inward, this notion of the American boy linked wildness with curiosity and authenticity" (*Designing the Creative Child*, xvi). Ogata discusses the book *Inventing Modern*, in which historian and mechanical engineer John H. Lienhard describes the imaginary of the savage boy inventor that was targeted in countless booklets and technical manuals going back to the 1860s, such as *The Boy Mechanic: 700 Things for Boys to Do* (1913). "The burgeoning world of *Modern* had systematically tutored us all in these rites of boyhood," Lienhard explains of these manuals (*Inventing Modern*, 192). He continues, "While there was nothing exceptional about *The Boy Mechanic*, it is a fine exemplar of the countless books whose explicit purpose was to shape boys during the early twentieth century" (194). Moreover, he shows that "boys were generally the sole focus on books like this": their titles, images, and text all reflected as much (196). Also see Chudacoff, *Children at Play*, 75.

91. Chudacoff, *Children at Play*, 95.

92. Toy manufacturers offered construction materials and electronics kits that appealed to parents' idealizations of childhood play and imagination and to their pedagogical goals for channeling it toward productive ends. Although many of these toys echoed adult roles and were seen by adults as a way to socialize children into future career paths, they claimed to do so through the natural, even "savage" inclinations that boys had toward certain kinds of play. Despite the fact that there was nothing inherent in engineering to make it masculine—it did not generally require great strength, for instance—these efforts continued a centuries-long association of science and learning with men. See also Abbiss, "Boys and Machines"; Alcorn, "Modeling Behavior"; Chudacoff, *Children at Play*; Cowan, "'Industrial Revolution' in the Home"; Douglas, *Listening In*; Ensmenger, "Making Programming Masculine"; Hicks, *Programmed Inequality*; Horowitz, *Boys and Their Toys?*; Ogata, *Designing the Creative Child*; and Oldenziel, "Boys and Their Toys."

93. Ruth Oldenziel makes the case that some companies, such as General Motors, specifically capitalized on this connection between technical hobbyists and the engineering careers they could later take up. The firm held a contest each year that was in fact a way to scout for talent. See Oldenziel, "Boys and Their Toys."

94. See e.g. See Abbiss, "Boys and Machines" and "Rethinking the 'Problem' of Gender"; Alper, "Can Our Kids Hack It?"; Ensmenger, "'Computer Boys' Take Over," *Computer Boys Take Over*, and "Signs of Rugged Individualism"; Hicks, "History of Computing" and *Programmed Inequality*; and Kleif and Faulkner, "Men's Pleasure in Technology."

95. Papert, *Children's Machine*, 4.

96. Oldenziel, "Boys and Their Toys."

97. Indeed, the category of "child" often does imply "boy" specifically, unless a product is specifically coded for girls (by being pink, for instance). Just as the category of "male" is often regarded as the default (against which "female" is the Other or is absent entirely), the category of "boy" is thus what the unmarked signifier "child" often represents, while "girl" is specifically marked and set apart. See e.g. Butler, *Gender Trouble*, chapter 1.

98. On online sexism, see Gray, "Intersecting Oppressions"; Kasumovic and Kuznekoff, "Insights into Sexism"; Kirkpatrick, "How Gaming Became Sexist"; Beavis and Charles, "'Real' Girl Gamer Please Stand Up"; Adam, "Cyberstalking and Internet Pornography"; Kelty, "Geeks, Social Imaginaries." On online racism, see Nakamura, *Cybertypes*; Nakamura and Chow-White, *Race after the Internet*; Fouché, "Say It Loud." On online harassment and exclusion, see Phillips, *Can't Have Nice Things*; Nagle, *Kill All Normies*; Citron, *Hate Crimes in Cyberspace*.

99. Turkle and Papert, "Epistemological Pluralism."

100. Ames and Rosner, "From Drills to Laptops."

101. Hall, "Encoding/Decoding."

102. Chudacoff, *Children at Play*, 43, 46.

103. See Blikstein, Seymour Papert tribute panel; Blikstein, "You Cannot Think about Thinking."

Chapter 2: Making the Charisma Machine

1. Bender et al., *Learning to Change the World*, 26.

2. This is based on longitudinal data collected through 2007 and 2008 on http://laptop.org/vision/people and http://wiki.laptop.org/go/Category:OLPC_People, as well as interviews with former OLPC employees in 2010–2014.

3. The mission statement was changed circa April 11, 2007. See Vota, "OLPC Mission Change."

4. Bender, "One Laptop per Child."

5. Krstić, "Google EngEDU Tech Talk."

6. Blizzard, "OLPC Analyst Meeting."

7. Negroponte, "No Lap Un-topped."

8. OLPC Wiki, "OLPC: Five Principles."

9. Coontz, *Way We Never Were.*

10. Boym, *Future of Nostalgia*, xiii–xiv.

11. Zuckerman, "It's Cute. It's Orange"; Edwards and Kim, "History of Processor Performance."

12. Bender et al., *Learning to Change the World*, 50.

13. Zuckerman, "It's Cute. It's Orange."

14. OLPC Wiki, "OLPC Myths."

15. Incidentally, the fan had adorned his XO with a label reading "Hey! Teacher! Leave us kids alone!" in reference to rock band Pink Floyd's *The Wall* but cast himself as one of the kids by changing "them" to "us." See Patel, "OLPC Hacked."

16. See, e.g., Norman and Draper, *User Centered System Design*; Namioka and Rao, "Introduction to Participatory Design"; Schuler and Namioka, *Participatory Design.*

17. One developer later told me that a closet full of laptops was the test of their networking capabilities, for instance.

18. Negroponte, "One Laptop per Child."

19. Zuckerman, "It's Cute. It's Orange."

20. See Bell and Dourish, "Yesterday's Tomorrows"; Dourish and Bell, "Resistance Is Futile"; Dourish and Bell, *Divining a Digital Future.*

21. Turner, *From Counterculture to Cyberculture*, 162–165.

22. Papert, *Mindstorms*, 10.

23. This is not to say that the OLPC community was completely monolithic in its vision for the machine. In particular, Ivan Krstić became disillusioned with the project, quit in early 2008 along with several others, and wrote a series of blog posts analyzing his own captivation with it and where it went astray. However, even as he was questioning some of the core tenets of OLPC, in particular its commitment to open-source software, he continued to believe that had it been done right, the project would have had the transformative effects that had convinced him to join initially. See, e.g., Krstić, "Maintaining Clarity"; Krstić, "This, Too, Shall Pass"; Krstić, "Sic Transit Gloria Laptopi"; Krstić, "Sweet Nonsense Omelet."

24. OLPC Wiki, "OLPC: Five Principles."

25. Levy, *Hackers*, 3–60; Brand, "Spacewar."

26. Papert, *Children's Machine*, 33.

27. An early OLPC employee told me in an interview that this bumpy texture was, oddly enough, meant to stop children from covering their laptops with stickers—not

so much to prevent children from customizing them as to prevent the disguising of stolen laptops. In my fieldwork, however, I still saw plenty of stickers adorning laptops, a practice that has since become widespread across the technology world as a way to personalize and distinguish laptops.

28. To help with the design, OLPC employed Fuseproject, an award-winning industrial design firm who had also worked with other technology companies, including Apple and Google. Prototyping was done by another industrial design firm, Design Continuum, based on a design competition that Google hosted in early 2005 (Bender et al., *Learning to Change the World*, 35).

29. Pentagram, "One Laptop per Child"; Landa, *Graphic Design Solutions*, 257.

30. "2B1" was a reference to a 1997 Media Lab conference and related foundation to bridge the digital divide, which involved both Negroponte and Papert but was spearheaded by Negroponte's son, Dimitri.

31. Though I distinguish fetishes from charismatic technologies, it is worth noting that this design has been fetishized—apart from the project's charismatic promises—by art and design collections. For instance, XO laptops are in the permanent collections of the Museum of Modern Art in New York and the San Francisco Museum of Modern Art.

32. Some in OLPC criticized this choice, including Ivan Krstić, who wrote, "The core mistake of the present Sugar approach is that it couples phenomenally powerful ideas about learning—that it should be shared, collaborative, peer to peer, and open—with the notion that these ideas must come presented in an entirely new graphical paradigm. We reject this coupling as untenable" (Krstić, "Sic Transit Gloria Laptopi").

33. OLPC Wiki, "Activity Pack."

34. One can install a game, or any XO software, by downloading the activity bundle (a file with the extension .xo) from a website, which will automatically install it, or by using the xo-get program manager on the XO, modeled on the Linux utility apt-get (see Hager, [Olpcaustria] "xo-get & svg-grabber"). For instance, to install the game SimCity, an XO user can navigate to the SimCity page on the OLPC wiki (http://wiki.laptop.org/go/SimCity) and click on the link to the activity bundle, or she can acquire the bundle in some other way that doesn't automatically install it—from a friend's laptop, for instance—and then type "xo-get install simcity.xo" in the terminal.

35. OLPC Wiki, "Games."

36. For a (somewhat gamer-centric) overview, see Gonzalez, "Two Tribes Go to War." For a discussion of the context of the controversy, see King and Borland, *Dungeons and Dreamers*.

37. Blizzard, "Doom on the XO."

38. OLPC Wiki, "Talk: Activities." Albert Cahalan is an open-source programmer who has contributed to various Linux projects, including several for OLPC such as Tux Paint (http://www.tuxpaint.org/download/xo).

39. Negroponte, *Being Digital*, 195.

40. Papert, *Children's Machine*, 4–5.

41. For instance, a game engine may include a physics engine that simulates the effects of gravity, inertia, and other physical forces on any object created in the game.

42. OLPC Wiki, "Game Jam Boston." Also see OLPC Wiki, "Game Jam."

43. When one XO laptop connects to another, each shows up on the other's Neighborhood view, along with access points and other technologies broadcasting in the 802.11 range. An XO laptop user can then connect with other XO laptops to move them to her Friends view, which is similar to the Neighborhood view but includes icons for only those laptops that are sharing activities with the user's XO (designated as Friends). From either of these views, a user can invite other XO users to any XO activity that is running, such as Browse or Chat (for more, see http://laptop.org/en/laptop/start/invite.shtml). A user can also make activities visible to various groups or to everyone in her neighborhood from the activity itself (see http://laptop.org/en/laptop/start/sharing.shtml). On the other end, a user who is invited to an activity can accept the invitation and then access the shared activity from her Neighborhood or Friends views.

44. Levy, *Hackers*, 31.

45. OLPC Wiki, "OLPC: Five Principles."

46. Papert, *Children's Machine, ix–x, 4, 15, 57–81*; Papert, *Mindstorms, x–xi, 67*.

47. Negroponte, "No Lap Un-topped."

48. Krstić, "Google EngEDU Tech Talk."

49. Zuckerman, "It's Cute. It's Orange."

50. Derndorfer, "OLPC School Servers."

51. OLPC Wiki, "Our Market: Black Market"; Fisher, "OLPC's XO Laptop."

52. Lydon, "One Laptop per Child?"

53. Blizzard, "One Laptop per Child."

54. Levy, *Hackers*, 28.

55. Regarding Logo as memory hog, see Thornburg, "Whatever Happened to Logo?," 83; Ames, "Hackers, Computers, and Cooperation."

56. Zuckerman, "It's Cute. It's Orange."

57. Papert, "Digital Development."

58. OLPC Wiki, "Disassembly"; OLPC Wiki, "Screws."

59. Many of the XO laptop's activities were written in Python, a programming language designed to be easy to learn, easy to read, and good for education, though also known to be slow. Other languages, ranging from ones designed for education to ones commonly used in production software, were also optionally included in the activity pack or could be downloaded from OLPC's websites.

60. Bender et al., *Learning to Change the World*, 60.

61. Hill, "Laptop Liberation."

62. Hill.

63. Hill, "Technological and Cultural Imperialism." For critiques of OLPC as culturally imperialist, see Fouché, "From Black Inventors"; Ananny and Winters, "Designing for Development"; Luyt, "Negotiation of Technological Meaning"; Chan, *Networking Peripheries*.

64. Papert, "Digital Development" (emphasis added).

65. Papert, *Mindstormss*, xx.

66. Hempel, "Miseducation of Young Nerds"; Thomas, *Hacker Culture*; Hill, "Geek Shall Inherit the Earth"; Negroponte, *Being Digital*.

67. In an analysis of the discourses underlying OLPC's interface guidelines in the context of computerization movements, researchers Mike Ananny and Niall Winters also critiqued the project's strongly individualistic model of social change. They found that OLPC assumed that social change would be initiated by children themselves and would progress through the computer-enabled connections between children. See Ananny and Winters, "Designing for Development." Also see Ames, "Hackers, Computers, and Cooperation."

68. OLPC Wiki, "OLPC Myths."

69. Negroponte, "Hundred Dollar Laptop." Note that although Negroponte did use this Trojan horse language in public talks such as this one, I could not find evidence that he actually said it to a head of state.

70. Negroponte, "Email Attachment." These three mailing lists were devel@laptop.org, sugar@laptop.org, and community-news@laptop.org.

71. OLPC Wiki, "OLPC: Five Principles."

72. OLPC Wiki, "OLPC Myths."

73. See Papert, *Mindstorms*, 28. Papert also states, "I have seen hundreds of elementary school children learn very easily to program, and evidence is accumulating to indicate that much younger children could do so as well" (13). However, note that he provides no citations or further information about this "evidence." Also see Papert, *Children's Machine*, 44, 50 ("School squanders its most valuable resource—the interchange between the most intellectually interesting students").

74. Papert, *Mindstorms*, 20–21.

75. Papert, *Mindstorms*, 56. Papert also notes that the Turtle "allows children to be deliberate and conscious in bringing a kind of learning with which they are comfortable and familiar to bear on math and physics. ... The Turtle in all its forms ... is able to play this role so well because it is both an engaging anthropomorphizable object and a powerful mathematical idea" (137; also see 56, 63, 129).
   Regarding conventional mathematics education, Papert writes,

> The children can see perfectly well that the teacher does not like math any more than they do and that the reason for doing it is simply that it has been inscribed into the curriculum. ... *I think it introduces a deep element of dishonesty into the educational relationship.*
>
> Children perceive the school's rhetoric about mathematics as double talk. In order to remedy the situation we must first acknowledge that the child's perception is fundamentally correct. The *kind of mathematics* foisted on children in schools is not meaningful, fun, or even very useful. ...
>
> This is the only mathematics they know. In order to break this vicious cycle I shall lead the reading into a new area of mathematics, Turtle geometry, that my colleagues and I have created as a better, more meaningful first area of formal mathematics for children. (50–51)

On this new approach to mathematics education, Papert writes,

> The computer-based Mathland I propose extends the kind of natural, Piagetian learning that accounts for children's learning a first language to learning mathematics. Piagetian learning is typically deeply embedded in other activities. For example, the infant does not have periods set aside for "learning talking." This model of learning stands in opposition to dissociated learning, learning that takes place in relative separation from other kinds of activities, mental and physical. In our culture. The teaching of mathematics in schools is paradigmatic of dissociated learning. ... In LOGO environments we have done some blurring of the boundaries: No particular computer activities are set aside as "learning mathematics." (48)

76. Papert, *Mindstorms*, 13.

77. Papert, *Mindstorms*, 21–22.

78. Papert, Mindstorms, 21–23, 105–109, 124–130, 173–174.; Papert, Children's Machine, 13.

79. OLPC Wiki, "OLPC Myths."

80. See examples throughout Papert, *Mindstorms*; Papert, *Children's Machine*.

81. Margolis and Fisher, *Unlocking the Clubhouse*, 15–93. On the term "subject position," see Hall, "Encoding/Decoding."

82. See Akrich, "De-scription of Technical Objects."

83. Oudshoorn, Rommes, and Stienstra, "Configuring the User."

84. See Ames, Rosner, and Erickson, "Worship, Faith, and Evangelism."

Chapter 3: Translating Charisma in Paraguay

1. Nickson, "General Election in Paraguay"; Barrionuevo, "Ex-cleric Wins Paraguay Presidency."

2. *Economist*, "Next Leftist on the Block."

3. *Economist*, "Boy and the Bishop"; *ABC Color*, "Lugo, padre y presidente."

4. Names of public figures, gleaned from public sources such as news articles, are not anonymized. This applies to most One Laptop per Child contributors, as well as the founders and high-level employees of Paraguay Educa, whose names I use with their permission. The town where Paraguay Educa's project was based is also not anonymized, as it has been publicly discussed and is easily findable. Other names—of teachers, students, and individual schools—are all anonymized or pseudonymized.

5. A month after Raúl's visit to OLPC in April 2007, Walter Bender visited Paraguay, and Raúl's university soon after received a donation of fifty experimental XO laptops.

6. *ABC Color*, "El proyecto Una laptop por niño."

7. Fried, "Negroponte."

8. Bender et al., *Learning to Change the World*.

9. Bender et al., *Learning to Change the World*; Derndorfer, "OLPC in Paraguay."

10. See Callon, "Sociology of Translation"; Callon, "Techno-economic Networks and Irreversibility."

11. Derndorfer, "4-Year Cost Increases"; Camfield, "Total Cost of XO Ownership"; Warschauer and Ames, "Save the World's Poor?," 36–37.

12. OLPC Staff. "OLPC Across the World: Rwanda." OLPC website. http://laptop.org/map/rwanda/.

13. Regions in Colombia, Guatemala, Brazil, Costa Rica, Mongolia, Nepal, Australia, and elsewhere tested the waters one city, district, or school at a time with small

projects of less than eight thousand laptops each. See Ames, "Translating Magic"; Warschauer and Ames, "Save the World's Poor?"

14. Chan, *Networking Peripheries*, 4–19, 96–108, 177–189.

15. See Negroponte, "Re-thinking Learning"; Papert and Freire, "Future of School." In an interview, Bender said, "Even go back to people like Dewey, Piaget, Paulo Freire; there are a number of people who have studied this and espouse learning through doing. And then you can take it and say, 'Well if you want more learning, you want more doing.' And it turns out that the laptop is a wonderful vehicle for doing. And so therefore there will be more learning" (Lydon, "One Laptop per Child?").

16. Chan, *Networking Peripheries*, 15–16, 125–151; Takhteyev, *Coding Places*, 9–10, 112–113; Vessuri, "Social Study of Science."

17. In the wake of OLPC's 2008 reorganization and downsizing, Daniel Drake visited Paraguay from February to June 2009 to help with the initial deployment. His blog posts about his experiences in Paraguay captured the idealism and excitement felt by not just him but the whole technical team in those early days. Drake left shortly after laptops were handed out, and his final report was filled with glowing accounts of the logistical achievements of the deployment and the excitement of the children with their brand-new laptops (see Drake, posts tagged "Paraguay"). Later, OLPC employee and volunteer Bernie Innocenti joined Paraguay Educa from January to August 2010 to help with ongoing software development and support (see Innocenti, "Brain Dump").

18. After their time with Paraguay Educa, Raúl worked in the United Kingdom and Martín studied in Spain. Raúl then moved to the San Francisco Bay Area to work at Facebook, Twitter, and Pinterest, and Martín returned to Asunción to contribute to OLPC's software, create robotics kits for Paraguay Educa's FabLab makerspace, mentor students through Google Summer of Code, and work remotely for San Francisco-based startup Endless Computers.

19. *ABC Color*, "El proyecto Una laptop por niño."

20. Caacupé was a strategic choice for Paraguay Educa's initial deployment. The town was close enough to Asunción to travel there and back in a day but far enough to remove concerns that it was just benefiting the relatively wealthier students in the capital city. Other municipalities that were interested were far to the north, along rough roads; Caacupé, by contrast, was along one of the main thoroughfares in Paraguay, linking Asunción with Ciudad del Este, where many imports (official and bootleg) arrived in the country. Moreover, Caacupé was familiar to at least some in the NGO because it hosted a huge religious festival every December that drew participants from the capital city, and some had second homes in the countryside nearby.

21. Drake, "XO Laptop Deployment Logistics."

22. This process was facilitated by the teachers, who were responsible for submitting lists of their students for XO laptop assignments. Municipal government officials made themselves available to register students before the laptop handout. See Paraguay Educa Wiki, "Boletin Trimestral."

23. Paraguay Educa Wiki, "Boletín Trimestral." One Argentinean commenter on Daniel Drake's blog explained that having a *cédula* in Argentina allows Argentinian citizens to vote and to access health services: "Those lazy parents [who don't register] make things incredibly hard for their children" (comment on Drake, "XO Laptop Deployment Logistics").

24. According to various international measures, corruption was mitigated (though not eradicated) during Lugo's presidency, and perceptions of corruption remain high. See Transparency International, "Paraguay"; Bureau of Democracy, Human Rights, and Labor, "Human Rights Practices: Paraguay"; Olhero and Sullivan, *Paraguay*. On a more personal note, I witnessed several shakedowns by corrupt policemen on the road between Asunción and Caacupé during my fieldwork. For a brief overview of atrocities during the Stroessner dictatorship, as discovered in the "terror archives" found in 1992, see Ceaser, "Paraguay's Archive of Terror."

25. *ABC Color*, "Inician entrega de 1.000 computadoras."

26. For a summary of what the municipal government wanted but was unable to do to help, see *ABC Color*, "La municipalidad jamás aportó." However, the government did come through with the funds to purchase one thousand laptops for incoming first-grade students in spring 2012 (see *ABC Color*, "Los alumnos del primer grado").

27. With deployments also underway in Uruguay and Peru, both run by their respective national governments, there was ample opportunity for regional solidarity through the XO laptop and also opportunities to base plans for a countrywide deployment in Paraguay on the experiences of these much-larger government-run programs—plans that Paraguay Educa tried to help along by proactively making contacts with those programs. See *ABC Color*, "El proyecto Una laptop por niño."

28. *ABC Color*, "Lugo dice que invertirá dinero"; *SciDev.Net*, "Paraguay ampliaría plan."

29. During my fieldwork in 2010, for instance, the Department of Education and Culture talked about its plans to install SMART Boards, or electronic whiteboards, in classrooms. Although nothing came of that project, it still took attention away from the project that Paraguay Educa was engaged in.

30. UNESCO Institute for Statistics, "Literacy Rate, Adult Total."

31. "Ceibal" is an acronym for Conectividad Educativa de Informática Básica para el Aprendizaje en Línea (Educational Connectivity via Basic Computing for Online Learning) and a reference to the ceibo tree, the national flower of Uruguay.

32. In fact, Paraguay's land ownership is one of the most uneven in the world, with 10% of the population owning 66% of the land and 30% of the population without land at all. See Gacitúa Marió, Silva-Leander, and Carter, *Paraguay: Social Development Issues*.

33. World Bank, *Paraguay: Estudio de pobreza*.

34. Federal Research Division, Library of Congress, "Country Profile: Paraguay." Though Paraguay does not provide an itemization of its education expenditures, a large proportion of this increase is likely spent on seven new public universities. Overall, the number of universities increased from one public and one private institution countrywide in 1989 to eight public and dozens of private institutions in 2010. See Central Intelligence Agency, "Paraguay."

35. WageIndicator Foundations, "Salarios mínimos en Paraguay."

36. Bureau of Democracy, Human Rights, and Labor, "Human Rights Practices: Paraguay."

37. "Infrastructure" has become a useful concept and analytical lens for a wide variety of systems that are invisible, embedded, relational, large scale, and "ready-to-hand" for those familiar with them, among other features (Star, "Ethnography of Infrastructure"; see also Larkin, "Politics and Poetics of Infrastructure"; Bowker et al., "Toward Information Infrastructure Studies"). The infrastructure that is most important in this account includes the power grid (especially as it extends into buildings via power outlets), the wireless internet networks installed by Paraguay Educa (WiMAX to Wi-Fi), the existing cellular networks, and the school buildings and furniture. We will see that all are prone to frequent visibility through breakdown.

38. WiMAX (Worldwide Interoperability for Microwave Access) is a wireless communications protocol well suited for longer-range transmission than more typical 802.11 Wi-Fi wireless local area network connections.

39. The frequent presence of Jesus pictures in schools is a nod to Paraguay's overwhelmingly Catholic majority, at nearly 90%, and in spite of the official separation of church and state in Paraguay.

40. Despite the proximity of the router, the internet connection was intermittent, possibly because of interference from the metal cage around the router and the rebar in the walls, or possibly because the school's bandwidth was overloaded. When I

interviewed him, one of Paraguay Educa's technicians scoffed at these metal cages: they cost more than the routers they housed and significantly reduced their signal strength by creating a partial Faraday cage around them, he explained. However, fears of theft were great enough that Paraguay Educa did not want to rely on hiding the routers inside light fixtures in classrooms or other obfuscation tactics.

41. Examples of such claims by Negroponte include:

"In China—I'm there every six weeks—they spend nineteen dollars per year per child on textbooks. So I say to the minister of education, it's a piece of cake. We will deliver this as a textbook with the textbooks in it. You amortize it over five years, and it's twenty dollars. Maybe it's twenty-two dollars if you want to charge some interest, or it's twenty-five dollars maybe. This can go through the textbook channel. In Brazil, the government pays for textbooks for all the kids, so they're already doing that. So it goes through the textbook channel. Textbooks are textbooks." (Negroponte, "Hundred Dollar Laptop")

"In most states (the United States is not one of them), most states control the textbooks, use the textbook channel. The Federal channel in Brazil does the textbooks. You may think of that as a liability but we use it as a feature, distribute through that channel." (Negroponte, "Digital, Life, Design")

"So when you look at a hundred dollar laptop, you amortize it over five years—let's pretend that's twenty dollars a year, no interest—Brazil and China both spend nineteen dollars per year per child in textbooks. So you could take the hundred dollar laptop, and economically justify it as an alternate textbook delivery plus these are updated textbooks plus it's any book, it's not just a selected few plus it's so-on and so forth." (Negroponte and Walsh, "Negroponte at NECC 2006")

"It gets distributed to the school system like a textbook." (Negroponte, "One Laptop per Child")

42. OLPC, "Vision: Mission." The quote continues, "They are tools to think with, sufficiently inexpensive to be used for work and play, drawing, writing, and mathematics." Negroponte used the same phrase in a number of interviews throughout 2005 and 2006.

43. Negroponte, "Hundred Dollar Laptop"; Lydon, "One Laptop per Child?"; Negroponte, "Email Attachment."

44. Zuckerman, "Child's Play." The OLPC Wiki states, "Early concept devices were shown with a hand crank on the side to demonstrate that they would work in areas where the only electricity available comes from devices like the Freecharge portable charger. This was removed due to concerns about stresses on the casing, and also ease of use" (OLPC Wiki, "XO Battery and Power").

45. OLPC employees brainstormed a long list of possible power sources for the XO on the project wiki, from the practical (car batteries, generators, solar panels, bikes) to the outlandish (cows, large boulders, handheld wind turbines—one contributor even referenced the failed PlayPump as potential inspiration). See OLPC Wiki, "XO Battery and Power."

46. OSILAC, *Households with ICTs.*

47. OLPC Wiki, "XO Battery and Power."

48. OLPC Wiki, "Principles and Basic Information"; Papert and Harel, "Situating Constructionism."

49. Paraguay was in a fairly unique, and quite lucky, position in regards to these crashes: Paraguay Educa's talented full-time software development team patched a significant number of bugs in the XO laptop's Sugar software package, many based on crash descriptions from the technical-support staff. For example, Paraguay Educa's developers disabled OLPC's much-vaunted "mesh network" capability when it failed to scale past half a dozen or so laptops and crashed all that were connected. Several times throughout my fieldwork, Paraguay Educa's technical staff circulated USB pen drives with new software distributions for children to install, and toward the end of my fieldwork, the software development team started to plan a system for pushing software updates to laptops over the network. Still, some bugs persisted, and new ones were sometimes inadvertently introduced, to the consternation of teachers. Moreover, relying on unreliable and often overloaded internet connections for fixes or installation files made recovery take even longer. Until they saw it firsthand, even some on Paraguay Educa's software development team thought that downloading and installing missing software should take only a few minutes at most, but this made two erroneous assumptions: that software installation was always bug free and that download speeds were always fast. Reality proved otherwise.

50. Bender et al., *Learning to Change the World*, xiii.

51. Cellan-Jones, "Negroponte—Throwing the Laptop"; Vota, "Negroponte Throws XO"; OLPC, "About the Laptop: Hardware."

52. Warschauer and Ames, "Save the World's Poor?"

53. OLPC Wiki, "Hardware Uniqueness"; OLPC, "About the Laptop: Hardware."

54. This shows that not all of these broken laptops were completely nonfunctioning; about one in three had more minor problems, perhaps with just a few broken keys, a broken charger, or dead pixels. One could borrow a charger or splice its cables for a temporary fix, tape bits of paper to the missing keys on the keyboard, or buy an external mouse if one's family was wealthy enough—and I did see some like this in use, especially in Caacupé's more wealthy private schools. However, the other two-thirds—about 15% of laptops overall—were not this minor. A broken screen was particularly problematic. It was not only common (afflicting 10% of laptops in August 2010), it rendered one's laptop unusable until a replacement screen was available. Only the charger broke more often, and only the motherboard cost more to replace.

55. The first-generation trackpad was especially bad: many who bought XO laptops through the Give One, Get One program complained about them. OLPC later updated the software to reset automatically when the trackpad started jittering. See OLPC Wiki, "XO-1 Touchpad: Issues."

56. Several of these problems were fixed in the second-generation XO (the XO-1.5), which Phase II of the deployment in Caacupé received in May 2011—though these machines had other problems, such as an unusually high wireless-card failure rate. And these second-generation machines did not help those in Phase I with first-generation XO laptops. Though many technologists may think nothing of replacing their devices frequently, hundreds of thousands of these first-generation XO machines are still in the hands of children around the world, with little hope or budget for an upgrade.

57. However, even with these resources, many broken laptops reportedly went unrepaired in Uruguay. See Derndorfer, "Plan Ceibal Expands"; Plan Ceibal, "Informe de evaluación"; Derndorfer, "4-Year Cost Increases"; Salamano et al., "Monitoreo y evaluación educativa."

58. Bender et al., *Learning to Change the World*, xiii.

59. These kinds of dynamics are corroborated by Toyama and his colleagues in India. See Pawar et al., "Multiple Mice for Retention Tasks"; Toyama, *Geek Heresy*, 4.

60. *ABC Color*, "El proyecto Una laptop por niño." In its first year, Plan Ceibal in Uruguay also conducted nationwide teacher training and debuted an OLPC-themed television show to introduce the XO laptop's activities to children and teachers alike.

61. NationMaster, "Paraguay: Media Stats"; Federal Research Division, Library of Congress, "Country Profile: Paraguay."

62. None at the time had smartphones. This was not long after the first iPhone had been released, and they were very expensive and hard to come by in Paraguay in 2008. I didn't see very many in 2010, either, but when I returned in 2013 they were ubiquitous.

63. Papert, *Mindstorms*, 6, 27, 206–207; Papert, *Children's Machine*, 4–5; Papert, "Digital Development"; Negroponte, Being Digital, 6, 2–4, 231.

64. Papert, *Children's Machines*, 137–153.

65. In full, she said, "I'd teach Phase II teachers to use activities without the internet, and I'd like them not on the internet for hours. The internet should only be a support; train teachers to use only the school server. Let the children learn the activities in Sugar—that would be my ideal. Sugar is educational. Use the XO like a [paper] notebook that has only limited use of the internet, because very few see

the internet's educational side. Having a foothold with the activities would be good."

66. See, for instance, Mosco, *Digital Sublime, 17–55*; Barlow, "Declaration of the Independence of Cyberspace"; Nissenbaum, "Contested Ontology of Cyberspace."

67. Interestingly, this conflicts with the results of an anonymous survey of Phase I teachers that I conducted in October 2010, in which all respondents who completed that portion of the survey (31 of the 48 total respondents, out of 148 teachers in Phase I) claimed to use the laptop in class at least once a week. Because of this conflict, I hypothesize a degree of social-desirability bias in answering this question.

68. I observed two schools in August 2009 and two in November 2010, all of them in the suburbs of the capital, Montevideo, within a ninety-minute bus ride from downtown.

69. See *ABC Color*, "Negocian, pero huelga de docentes continúa"; *ABC Color*, "Diez gremios de educadores."

70. During and after my fieldwork, teachers' unions sought a 22% raise to their wages. See *ABC Color*, "Negocian, pero huelga de docentes continúa."

71. Vota, "XO Helicopter Deployments?"; Hachman, "Negroponte."

72. See Oppenheimer, "Latin America Leads"; Warschauer and Ames, "Save the World's Poor?"; *El Observador*, "La ceibalita un millón."

73. Peru is a mountainous country with a large rural indigenous population (numbering approximately thirteen and a half million of Peru's thirty million inhabitants), which has historically put up strong resistance to the colonialist policies of the elite in the capital city, Lima. This made Peru a challenging environment for a government-led educational technology program from the start. Between 2007 and 2010, Peru's national government, headed up by president Alan García, gave laptops to some 450,000 of the approximately 4 million schoolchildren around the country, focusing largely on rural areas. A new administration headed by president Ollanta Humala purchased nearly 500,000 additional laptops for equipping shared computer labs in schools across Peru in 2011 and 2012. See Derndorfer, "Interview with Sandro Marcone."

74. In a 2010 interview, Hernan Pachas, a technical lead for OLPC Peru and a vocal proponent of Peru's project on OLPC mailing lists, told me that the power problem would be corrected with solar panels, but no solar panels were ever distributed. See also Derndorfer, "OLPC in Peru"; Breitkopf, "Observations of 'Una laptop por nino'"; Patzer, "Who's to Blame?"

75. *Economist*, "Error Message"; Santiago et al., *Experimental Assessment*; Cristia et al., "Technology and Child Development."

76. Derndorfer, "OLPC in Peru"; Patzer, "Who's to Blame?"; Breitkopf, "Observations of 'Una laptop por nino.'"

77. Anita Say Chan's book *Networking Peripheries* chronicles the troubles of OLPC Peru, but also charts a potential rebirth. Chan describes a local movement underway in 2012 to resurrect a small part of this troubled project, spearheaded by local activists in Puno and by free software advocates from across Peru and the world. This group worked to reappropriate OLPC's XO laptops as tools for fostering activism and local pride by translating the interface into Peru's two most spoken indigenous languages (Quechua and Aymara) and developing guides for local teachers. Although I haven't heard reports that this initiative ultimately had much effect on teachers or students, the innovation that Chan witnessed was fostered by the charismatic promises of both the XO laptop and free/open-source software. See Chan, "Hacking Digital Universalism"; Chan, *Networking Peripheries*, 173–196.

78. Salamano et al., "Monitoreo y evaluación educativa"; Martínez, Alonso, and Díaz, "Monitoreo y evaluación." This message of popularity was echoed at a self-congratulatory conference hosted by Plan Ceibal in November 2010, which I attended (and at which Uruguay's newly elected president, José Mujica, gave a keynote talk). See Montevideo.com, "Sobre el cielo austral." Similar results were found by preliminary evaluations of much smaller pilot programs elsewhere, where one classroom or school was given laptops with a hope of demonstrating a proof of concept or establishing a successful set of practices and later expanding the program. Many of these did not later expand. See Lowes and Luhr, *Evaluation of the Teaching Matters*; ACER, *Evaluation of One Laptop per Child*; Andersen, "Lines of Marginalisation"; Andersen, "Travelogue of 100 Laptops."

79. Melo et al., "Profundizando en los efectos." These findings echoed the mixed results of an educational assessment I helped Paraguay Educa run during my fieldwork in Paraguay (see Appendix A).

80. Negroponte also neglected to consider the colonialist/apartheidist perspective of the movie. See Lee, "Gods Must Be Crazy"; Gugler, "Critic's Responsibility."

Chapter 4: Little Toys, Media Machines, and the Limits of Charisma

1. Papert, *Mindstorms*, viii.

2. OLPC Wiki, "Disassembly"; OLPC Wiki, "Screws."

3. Bullis, "Hundred-Dollar Laptop."

4. PlayStation game consoles were not common, but I did encounter them in about 5% of the households I visited, and 5% of teachers reported having one in their house in a survey I conducted.

5. The parents in most of the households that had computers were teachers, politicians, or professionals. Via a survey that I conducted with teachers during my fieldwork, I found that just over one-quarter of teachers in schools with XO laptops had another computer at home, and 13% had internet. The incidence of both computers and the internet was much lower among non-white-collar workers.

6. It was, in fact, possible to connect the XO to the internet via USB stick, but it did require insider knowledge about the XO and a high degree of technical expertise that only Paraguay Educa programmers and technical staff had.

7. NationMaster, "Paraguay: Media Stats."

8. The concerns that parents had about computers and the internet included obesity, poor eyesight, predators, pornography, and violent content.

9. Papert, *Mindstorms*, viii.

10. These numbers narrowed somewhat for hardware problems that remained unfixed by Paraguay Educa's technicians: per capita, there were 2.15 unfixed laptops in rural areas for every unfixed laptop in urban areas, according to my August 2010 Inventario snapshot.

11. "Wine" is a recursive "backronym" for "Wine Is Not an Emulator."

12. Anita Say Chan analyzes imagery of this form used by OLPC Peru. See Chan, "Hacking Digital Universalism"; Chan, *Networking Peripheries*, 180–181.

13. For supporting views, see Jenkins, *Confronting the Challenges*; Ito et al., *Connected Learning*; Ito et al., *Hanging Out, Messing Around*. For critiques, see Warschauer and Matuchniak, "New Technology and Digital Worlds"; Sims, "Video Game Culture"; Sims, *Disruptive Fixation*.

14. For the OLPC view, see Papert, "Digital Development"; Bender, "One Laptop per Child"; Bender et al., *Learning to Change the World*; Negroponte, "No Lap Untopped." For critiques, see Ananny, "$100 Laptop"; Ananny and Winters, "Designing for Development"; Luyt, "Negotiation of Technological Meaning"; Vota, "Only Hope to Eliminate Poverty."

15. Negroponte, *Being Digital*, 204.

16. Papert, *Children's Machine*, 4.

17. See, e.g., Wang et al., "Action Video Game Training."

18. Westervelt, "Cubist Revolution"; Ames and Burrell, "'Connected Learning' and the Equity Agenda."

19. For a detailed analysis of the culture of the Sony Walkman, see Gay et al., *Doing Cultural Studies*.

20. Internet access at home was rare in 2010. Even among teachers, who tended to be more technologically inclined than the general population, only 13% had home internet connections.

21. Papert, "Digital Development."

22. Vota, "One Pornographic Image."

23. See, e.g., American Academy of Pediatrics, "Media and Children Communication Toolkit."

24. See, e.g., Electronic Frontier Foundation, "Content Blocking."

25. Papert, Mindstorms, vii, 7, 9, 19, 40, 50, 96, 114–115, 174; Papert, *Children's Machine*, ix, 1–5, 13, 140; Papert et al., *Logo Philosophy and Implementation*, ix.

26. Blocking YouTube was a difficult decision, one teacher explained, but necessary because of children watching what she described as "aggressive" videos. "It is too bad for other kids who are not interested in those things," she lamented. This suggests an interesting cultural difference, which we will return to below: whereas YouTube aggressively filters sexual content to make it "kid safe" according to cultural values in the United States, the cultural values in Paraguay condemned violence where US cultural values are more equivocal on the topic.

27. Blizzard, "Doom on the XO"; Murph, "Caught Playing Super Mario."

28. "Pen drive," along with "laptop" and "InFocus" (for a projector), were among a number of words borrowed from English to describe technology that I encountered during my fieldwork.

29. Resnick et al., "Scratch: Programming for All."

30. Fouché, "From Black Inventors." Also see Luyt, "Negotiation of Technological Meaning"; Ananny and Winters, "Designing for Development"; Toyama, "Can Technology End Poverty?"

31. See, e.g., Bender, "One Laptop per Child"; Bender et al., *Learning to Change the World*; Hempel, "Miseducation of Young Nerds"; Hill, "Geek Shall Inherit the Earth," "Laptop Liberation," and "Technological and Cultural Imperialism"; Negroponte, "No Lap Un-topped," "Hundred-Dollar Laptop," and *Being Digital*; OLPC Wiki, "OLPC: Five Principles" and "OLPC Myths"; and Papert, "Digital Development."

32. Amico, "Nickelodeon Partners with OLPC."

33. *La Nación*, "Alumna del cuarto grado."

34. See Fejes, "Media Imperialism."

35. See Wolf, *Europe*.

36. See Parker, "Closing the Digital Divide"; Wilson, Wallin, and Reiser, "Social Stratification." For critiques, see Warschauer, "Demystifying the Digital Divide"; Richtel, "Wasting Time."

Chapter 5: The Learning Machine and Charisma's Cruel Optimism

1. Papert, *Mindstorms*, 28, 44, 50; OLPC Wiki, "OLPC Myths."

2. Negroponte, "Email Attachment."

3. Papert, *Mindstorms*, 37.

4. Papert, "Digital Development."

5. Papert, *Mindstorms*, 6, 16.

6. Papert, *Mindstorms*, 20, 156. This critique also applies to Papert's mentor, Jean Piaget, who first developed his theories based on observations of his own three children yet did not acknowledge the role that their upper-middle-class Swiss upbringing played in shaping them.

7. Papert, *Mindstorms*, 9, 11.

8. For more on social worlds, see Becker, *Art Worlds*.

9. See Lave and Wenger, *Situated Learning*.

10. Resnick et al., "Scratch: Programming for All."

11. See Nestlé Uruguay, "Vascolet."

12. See Ito et al., *Connected Learning*.

13. NationMaster, "Paraguay: Media Stats"; Federal Research Division, Library of Congress, "Country Profile: Paraguay."

14. Brennan, "Scratch@MIT 2010."

15. I later learned that the broader OLPC community had found a clunky workaround. See OLPC Wiki, "Video Chat."

16. See Lave and Wenger, *Situated Learning*; Barron et al., "Parents as Learning Partners"; Ito, *Engineering Play*; Ito et al., *Living and Learning*; Lave and McDermott, "Estranged Labor Learning"; Lareau, "Social Class Differences"; Lareau and Shumar, "Problem of Individualism"; Brown et al., "Situated Cognition."

17. Negroponte, "Negroponte at WEF."

18. Takhteyev, *Coding Places*, 26–29; Takhteyev, "Networks of Practice," 566–567; Lave and Wenger, *Situated Learning*.

19. Chan, "Hacking Digital Universalism"; Chan, *Networking Peripheries*, 180–181.

20. Ferguson, *Global Shadows*, 19–20.

21. The proportion of English on the internet, while falling, still accounts for over half of content and is at least an order of magnitude more than any other language. W3Techs, "Usage of Content Languages."

22. Appadurai, "Global Ethnoscapes"; Appadurai, *Modernity at Large*; Berlant, *Cruel Optimism*.

Chapter 6: Performing Development

1. See Warschauer and Matuchniak, "New Technology and Digital Worlds"; Vigdor, Ladd, and Martinez, "Scaling the Digital Divide"; Malamud and Pop-Eleches, "Home Computer Use."

2. Papert, *Children's Machine*, 44.

3. Some schools had a morning session from seven to eleven, but this one, which drew children from rural surroundings, started at eight to allow children more time to get to school.

4. Ames, "Hackers, Computers, and Cooperation," 18:11.

5. Bender et al., *Learning to Change the World*, 61.

6. Bender et al., *Learning to Change the World*, 113.

7. Bender et al., *Learning to Change the World*, 1.

8. Bender et al., *Learning to Change the World*, 128.

9. Weber, "Charismatic Authority."

10. Goffman, *Presentation of Self in Everyday Life*s, 8–46, 78.

11. See Butler, *Excitable Speech*.

12. For instance, Lilly Irani has described hackathons as performances, drawing on an ethnography of a civic hackathon in India (Irani, "Making of Entrepreneurial Citizenship"). Jenna Burrell and Elisa Oreglia have deconstructed the "performance" that the idea of mobile-phone access does in development circles, where they supposedly enable access to market prices and other information for farmers, despite no evidence of or need for such (Burrell and Oreglia, "Myth of Market Price Information"; Srinivasan and Burrell, "Fishers of Kerala"). James Ferguson has explored the way that technology access can also perform to those being "developed," examining the hopes of a group of upper-class technophiles in Zambia (Ferguson, *Global Shadows*, esp. chap. 5). A few scholars have referenced "performing development" specifically: Saida Hodzic has shown how women's NGOs "perform development"

to donors in Ghana (Hodzic, *Performing Development*); John Lauermann has discussed the economic performativity of the sale of the Yemeni street drug qat (Lauermann, "Performing Development"); and John Clammer has discussed the role of theater performances in shaping development discourses (Clammer, "Performing Development").

13. Twist, "Debut for $100 Laptop."

14. Cellan-Jones, "Negroponte—Throwing the Laptop"; Vota, "Negroponte Throws XO."

15. Thompson, "Negroponte Plans Tablet Airdrops"; Hachman, "Negroponte."

16. Negroponte, "30-Year History."

17. Talbot, "Tablets but No Teachers"; Condliffe, "Ethiopian Kids Hack"; Thompson, "African Kids Learn"; Keller, Smith, and McNierney, "Reading Project."

18. Hassan, "MIT Media Lab"; Brand, *Media Lab*, 3–16.

19. Among others, sociologist Michael Burawoy does explore the differences between positivist and reflexive research methods, arguing that they are and should be held to somewhat different standards—but that both can be just as rigorous in their interpretations on the world. See Burawoy, "Extended Case Method." In contrast, Papert leans on "ethnography" as an excuse for presenting what he admits is "anecdotal" evidence, but without engaging in any of the analytic rigor that has long been a standard in ethnographic methods. In a 1987 essay he stated, "One often hears that reports of good Logo environments are 'anecdotal.' This word is used as a derogatory form of the adjective 'ethnographic' and in contrast to a more 'scientific method.' I do not agree with the derogation of the case study approach." Papert, "Computer Criticism," 26.

20. Papert, *Mindstorms*, viii; Papert, *Children's Machine*, 21, 26–27, 196; Papert, "Computer Criticism"; Papert, "Critique of Technocentrism."

21. See Dray and Menosky, "New World Order"; Camfield, "OLPC History"; Papert et al., *Logo Philosophy and Implementation*, ix–x, 2–21; Papert, *Children's Machine*, 75–78. Warschauer, *Laptops and Literacy*.

22. See, e.g., Vota, "How to ... Use Failures."

23. Asghar, "Silicon Valley's 'Fail Fast' Mantra."

24. Fail Festival, "Why."

25. MacKenzie, *Engine, Not a Camera*.

26. Mazzarella, "Beautiful Balloon," 799.

27. Mazzarella, "Beautiful Balloon," 799.

28. Mazzarella, "Beautiful Balloon," 784.

29. Dourish and Bell, *Divining a Digital Future*, 3–5; Mosco, *Digital Sublime*, 3–4, 13–14, 24, 28, 31, 34–37, 41–42, 68, 74–75, 85, 88, 94–97, 118, 125.

30. Ames, Rosner, and Erickson, "Worship, Faith, and Evangelism."

Chapter 7: Conclusion

1. Warren, "Chromebooks Outsold Macs."

2. Talbot, "Tablets but No Teachers."

3. Keller, Smith, and McNierney, "Reading Project"; Talbot, "Tablets but No Teachers"; Condliffe, "Ethiopian Kids Hack"; Thompson, "African Kids Learn." See also chapter 6 for a more detailed description of how this Ethiopia project was run and what it accomplished.

4. Barshay, Jill. "Getting boys—and girls—interested in computer coding." *Education by the Numbers* (blog), August 12, 2014. http://educationbythenumbers.org/content/girls-computer-coding_1691/.

5. Tyack and Cuban, *Tinkering toward Utopia,* 1–11, 110–111; Cuban, *Teachers and Machines*, 4–5; Cuban, Oversold and Underused, 1, 179, 195–197.

6. Khan, "Let's Use Video"; Kaplan, "Sal Khan."

7. See, e.g., Clow, "Funnel of Participation"; Liyanagunawardena, Adams, and Williams, "MOOCs"; Paul, "MOOC Gender Gap"; Christensen et al., "MOOC Phenomenon"; Khalil and Ebner, "MOOCs Completion Rates"; Dillahunt, Wang, and Teasley, "Democratizing Higher Education"; Engle, Mankoff, and Carbrey, "Human Physiology Course"; Ahearn, "Abysmal MOOC Completion Rates."

8. A charter school is an alternative school environment that is accountable to parents but not to elected school boards. Charter schools, moreover, can select which students enroll, meaning that students deemed high risk in some way (students with disabilities, for instance, and sometimes English-language learners) can be barred from attending.

9. Papert, *Mindstorms*, 37; Papert, *Children's Machine*, 8–9, 12–13; Sullivan, "Man behind Logo."

10. See Sullivan, "Man behind Logo"; Dray and Menosky, "New World Order."

11. For examples of this trend, see Stack Overflow, "2015 Developer Survey"; Biba and McKenna, "Turn On, Code In, Drop Out"; *TNW*, "Half of Developers."

12. The fellowship website states, "The Thiel Fellowship gives $100,000 to young people who want to build new things instead of sitting in a classroom."

Applicants must be 22 years old or younger. As of July 2018, 124 fellowships had been awarded. See http://thielfellowship.org/; Azoulay et al., "Age and High-Growth Entrepreneurship."

13. OLPC News, "Criticising OLPC." Negroponte also mentioned the Red Cross parallel in several other talks and interviews. See Negroponte, "More Cause than Opportunity"; Negroponte, "Digital, Life, Design"; Negroponte, "Hundred Dollar Laptop."

14. See Ames, Rosner, and Erickson, "Worship, Faith, and Evangelism."

15. For a description of ideological work, see Berger, *Survival of a Counterculture*, 18–22.

16. Oudshoorn, Rommes, and Stienstra, "Configuring the User," 53–55.

17. Mosco, *Digital Sublime*, 4, 29, 41, 142.

18. Mosco, *Digital Sublime*, 55–84.

19. See Nye, *American Technological Sublime*; Douglas, *Listening In*; Mosco, *Digital Sublime*; Haring, *Ham Radio's Technical Culture*.

20. Winner, "Technology Today," 1000-1001.

21. See Mosco, *Digital Sublime*; Turner, *From Counterculture to Cyberculture*.

22. Mosco, *Digital Sublime*, 112–113; Turner, "Burning Man at Google"; Turner, "Counterculture Met the New Economy"; Turner, *From Counterculture to Cyberculture*.

23. Ames and Rosner, "From Drills to Laptops."

24. Segal, *Technological Utopianism*, 170.

25. Mosco, *Digital Sublime*, 22.

26. Mosco, *Digital Sublime.*, 3–6.

27. Tyack and Cuban, *Tinkering toward Utopia*, 1–11; Ames, "Charismatic Technology."

28. Negroponte, "Laptops Work."

29. See McLuhan, *Understanding Media*.

30. See, e.g., Winner, "Opening the Black Box"; Law, "Introduction"; Akrich, "Description of Technical Objects."

31. Toyama's *Geek Heresy* provides many such examples.

32. See Torn Halves, "Sugata Mitra"; Clark, "Mitra's 'Hole-in-Wall' Project"; Clark, "Sugata Mitra: Slum Chic?"; Bennett, "Sugata Mitra"; Wilby, "Sugata Mitra"; Cappelle, Evers, and Mitra, "Unsupervised Computer Use"; Mitra et al., "Acquisition

of Computing Literacy"; Mitra, "Minimally Invasive Education"; Cuban, "No End to Magical Thinking"; Nava, "Some Obvious Notes"; Mitra, *Hole in the Wall*; Ward, "Critique of Hole-in-the-Wall."

33. Geertz, "Thick Description," 5.

34. Some have, indeed, wondered whether Nicholas Negroponte might have learned some bad lessons from his older brother, John, who has been criticized for his willingness to interfere in Latin American democratic processes on behalf of the United States. For more on John Negroponte, see Shane, "Central Negroponte Role"; Gutman, *Banana Diplomacy*.

35. See, e.g., Ananny and Winters, "Designing for Development"; Luyt, "Negotiation of Technological Meaning"; Fouché, "From Black Inventors."

36. See Scheper-Hughes, *Death without Weeping*; Finkle, "Nestlé, Infant Formula, and Excuses"; Olivera and Lewis, ¡Cochabamba!

37. See Medina, Marques, and Holmes, *Beyond Imported Magic*; Medina, *Cybernetic Revolutionaries*; Chan, *Networking Peripheries*; Takhteyev, *Coding Places*; Vessuri, "Social Study of Science"; Alvarez, Dagnino, and Escobar, *Cultures of Politics, Politics of Cultures*.

38. One of the best known is the utopian vision of a technology-rich cybernetic socialist society in 1970s Chile, before president Salvador Allende was overthrown and assassinated in 1973. Eden Medina provides a detailed account of the negotiations between local actors and international "experts" over the implementation of this Marxist cybernetic vision (See Medina, *Cybernetic Revolutionaries*). This bears resemblance to OLPC projects across Latin America in how negotiations played out between the utopian visions put forth by the powerful but disconnected developers and consultants, the governments or NGOs on the ground trying to implement and adapt these visions, and the beneficiaries enrolled in this vision by these powerful actors.

39. Ferguson, *Global Shadows*, 89–112.

40. See Chan, "Coding Free Software"; Chan, "Hacking Digital Universalism"; Chan, *Networking Peripheries*.

41. Papert, "Digital Development"; Papert et al., *Logo Philosophy and Implementation*, 133.

42. Ybema and Kamsteeg, "Making the Familiar Strange," 102, 109, 116. See also Mills, *Sociological Imagination*.

43. Dobrin, *Constructing Knowledges*, 123.

44. Ames, "Translating Magic," 394; Ames, "Charismatic Technology," 117–118.

45. Tyack and Cuban, *Tinkering toward Utopia*, 5.

46. Haraway, *Staying with the Trouble*, 2.

47. See Hall, "Rediscovery of 'Ideology.'"

Appendix A: An Assessment of Paraguay Educa's OLPC Project

1. The disproportionate increase in the numbers of students in Phase I schools—with 47% more students in third and sixth grades in 2013 than in 2010, compared to 18% more in Phase II schools and 10% more in the control schools—was itself a surprising finding and one for which we do not have a nonspeculative explanation. Some of this increase may be due to reduced dropouts or increased school attendance in Phase I schools or, perhaps, to just more people moving to the town center.

2. These blank tests could have been from a student who did not fill anything out or from a student who was absent on the main test day as well as the makeup test days but still had a paper test put aside by the proctor. Blank tests might also have been inconsistently recorded by those entering data, especially between 2010 and 2013. Therefore, I excluded all of them for consistency.

3. We chose this method of analysis because it provides a correction for when the samples in question are not guaranteed to have equal variances.

Appendix B: Methods for Studying the Charisma Machine

1. Geertz, "Thick Description."

Bibliography

Abbiss, Jane. "Boys and Machines: Gendered Computer Identities, Regulation and Resistance." *Gender and Education* 23, no. 5 (2011): 601–617. https://doi.org/10.1080/09540253.2010.549108.

———. "Rethinking the 'Problem' of Gender and IT Schooling: Discourses in Literature." *Gender and Education* 20, no. 2 (2008): 153–165. https://doi.org/10.1080/09540250701805839.

ABC Color. "Diez gremios de educadores van a huelga en busca del salario mínimo." September 21, 2011. http://www.abc.com.py/edicion-impresa/locales/diez-gremios-de-educadores-van-a-huelga-en-busca-del-salario-minimo-310662.html.

———. "El proyecto Una laptop por niño comenzará a funcionar en Paraguay." September 25, 2008. http://www.abc.com.py/edicion-impresa/locales/el-proyecto-una-laptop-por-nino-comenzara-a-funcionar-en-paraguay-1105257.html.

———. "Inician entrega de 1.000 computadoras en Caacupé." March 19, 2012. http://www.abc.com.py/nacionales/inician-entrega-de-1000-computadoras-en-caacupe-380849.html.

———. "La municipalidad jamás aportó para que Caacupé sea una ciudad digital." March 2, 2012. http://www.abc.com.py/edicion-impresa/interior/la-municipalidad-jamas-aporto-para-que-caacupe-sea-una-ciudad-digital-375149.html.

———. "Los alumnos del primer grado de Caacupé ya recibieron sus laptops." March 24, 2012. http://www.abc.com.py/edicion-impresa/locales/los-alumnos-del-primer--grado-de-caacupe--ya-recibieron-sus-laptops-382581.html.

———. "Lugo dice que invertirá dinero de Itaipú en computadoras para niños." April 30, 2011. http://www.abc.com.py/edicion-impresa/politica/lugo-dice-que-invertira-dinero-de-itaipu-en-computadoras-para-ninos-251363.html.

———. "Lugo, padre y presidente." April 22, 2009. http://www.abc.com.py/edicion-impresa/politica/lugo-padre-y-presidente-1166109.html.

———. "Negocian, pero huelga de docentes continúa." July 12, 2010. http://www.abc.com.py/edicion-impresa/locales/negocian-pero-huelga-de-docentes-continua-130246.html.

ACER (Australian Council for Educational Research). *Evaluation of One Laptop per Child (OLPC) Trial Project in the Solomon Islands*. Ministry of Education and Human Resources Development, Solomon Islands Government, March 2010. https://research.acer.edu.au/digital_learning/7/.

Adam, Alison. "Cyberstalking and Internet Pornography: Gender and the Gaze." *Ethics and Information Technology* 4, no. 2 (June 2002): 133–142. https://doi.org/10.1023/A:1019967504762.

Ahearn, Amy. "The Flip Side of Abysmal MOOC Completion Rates? Discovering the Most Tenacious Learners." *EdSurge*, February 22, 2017. https://www.edsurge.com/news/2017-02-22-the-flip-side-of-abysmal-mooc-completion-rates-discovering-the-most-tenacious-learners.

Akrich, Madeleine. "The De-scription of Technical Objects." In *Shaping Technology/Building Society: Studies in Sociotechnical Change*, edited by Wiebe E. Bijker and John Law, 205–224. Cambridge, MA: MIT Press, 1992.

Alcorn, Aaron L. "Modeling Behavior: Boyhood, Engineering, and the Model Airplane in American Culture." PhD diss., Case Western Reserve University, 2009. OhioLINK (case1220640446).

Alper, Meryl. "'Can Our Kids Hack It with Computers?': Constructing Youth Hackers in Family Computing Magazines (1983–1987)." *International Journal of Communication* 8 (2014): 673–698. https://ijoc.org/index.php/ijoc/article/view/2402.

Alvarez, Sonia E., Evelina Dagnino, and Arturo Escobar. *Cultures of Politics, Politics of Cultures: Re-visioning Latin American Social Movements*. Boulder, CO: Westview, 1998.

American Academy of Pediatrics. "Media and Children Communication Toolkit." Accessed November 6, 2018. https://www.aap.org/en-us/advocacy-and-policy/aap-health-initiatives/Pages/Media-and-Children.aspx.

Ames, Morgan G. "Charismatic Technology." In *Proceedings of the Fifth Decennial Aarhus Conference on Critical Alternatives*, chaired by Olav W. Bertelsen, Kim Halskov, Shaowen Bardzell, and Ole Iversen, 109–120. N.p.: Aarhus University Press, 2015. https://doi.org/10.7146/aahcc.v1i1.21199.

———. "Hackers, Computers, and Cooperation: A Critical History of Logo and Constructionist Learning." *Proceedings of the ACM on Human-Computer Interaction* 2, no. CSCW (2018). https://doi.org/10.1145/3274287.

———. "Managing Mobile Multitasking: The Culture of iPhones on Stanford Campus." In *Proceedings of the 2013 Conference on Computer Supported Cooperative*

Work, chaired by Amy Bruckman, Scott Counts, Cliff Lampe, and Loren Terveen, 1487–1498. New York: ACM, 2013. https://doi.org/10.1145/2441776.2441945.

———. "Translating Magic: The Charisma of One Laptop per Child's XO Laptop in Paraguay." In *Beyond Imported Magic: Essays on Science, Technology, and Society in Latin America*, edited by Eden Medina, Ivan da Costa Marques, and Christina Holmes, 207–224. Cambridge, MA: MIT Press, 2014.

Ames, Morgan G., and Jenna Burrell. "'Connected Learning' and the Equity Agenda: A Microsociology of Minecraft Play." In *Proceedings of the 2017 ACM Conference on Computer Supported Cooperative Work and Social Computing*, chaired by Charlotte P. Lee, Steve Poltrock, Louise Barkhuus, Marcos Borges, and Wendy Kellogg, 446–457. New York: ACM, 2017. https://doi.org/10.1145/2998181.2998318.

Ames, Morgan G., and Daniela K. Rosner. "From Drills to Laptops: Designing Modern Childhood Imaginaries." *Information, Communication & Society* 17, no. 3 (2014): 357–370. https://doi.org/10.1080/1369118X.2013.873067.

Ames, Morgan G., Daniela K. Rosner, and Ingrid Erickson. "Worship, Faith, and Evangelism: Religion as an Ideological Lens for Engineering Worlds." In *Proceedings of the 18th ACM Conference on Computer Supported Cooperative Work & Social Computing*, chaired by Dan Cosley, Andrea Forte, Luigina Ciolfi, and David McDonald, 69–81. New York: ACM, 2015. https://doi.org/10.1145/2675133.2675282.

Amico, Giulia. "Nickelodeon Partners with OLPC on Multimedia Contest." One Laptop per Child blog, July 15, 2011. https://web.archive.org/web/20151212071530/http://blog.laptop.org/2011/07/15/nickelodeon-olpc-contest/

Ananny, Mike. "The $100 Laptop: Trojan Horse, Piñata or Social Movement?" Unpublished paper (provided by author), December 13, 2005.

Ananny, Mike, and Niall Winters. "Designing for Development: Understanding One Laptop per Child in Its Historical Context." In *2007 International Conference on Information and Communication Technologies and Development*. N.p.: IEEE, 2007. https://doi.org/10.1109/ICTD.2007.4937397.

Andersen, Lars Bo. "Lines of Marginalisation in a One Laptop per Child Project." Paper presented at the Society for Social Studies of Science (4S)/European Association for the Study of Science and Technology (EASST) Joint Conference, Copenhagen Business School, Copenhagen, Denmark, October 19, 2012. http://www.larsbo.org/publications/olpc/lines-of-marginalisation-in-an-one-laptop-per-child-project.

———. "A Travelogue of 100 Laptops: Investigating Development, Actor–Network Theory and One Laptop per Child." PhD diss., Aarhus University, 2013. http://www.laptopstudy.net/.

Anderson, Benedict. *Imagined Communities: Reflections on the Origin and Spread of Nationalism*. London: Verso, 1983.

Appadurai, Arjun. "Global Ethnoscapes." In *Recapturing Anthropology: Working in the Present*, edited by Richard G. Fox, 191–210. Santa Fe, NM: School of American Research Press, 1991.

———. *Modernity at Large: Cultural Dimension of Globalization*. Minneapolis: University of Minnesota Press, 1996.

Asghar, Rob. "Why Silicon Valley's 'Fail Fast' Mantra Is Just Hype." *Forbes*, July 14, 2014. https://www.forbes.com/sites/robasghar/2014/07/14/why-silicon-valleys-fail-fast -mantra-is-just-hype/.

Azoulay, Pierre, Benjamin F. Jones, J. Daniel Kim, and Javier Miranda. "Age and High-Growth Entrepreneurship." NBER Working Paper No. w24489, April 2018. https://ssrn.com/abstract=3158929.

Badham, John, dir. *WarGames*. MGM/UA Entertainment, 1983.

Barlow, John Perry. "A Declaration of the Independence of Cyberspace." February 8, 1996. Electronic Frontier Foundation. https://www.eff.org/cyberspace-independence.

Barrionuevo, Alexei. "Ex-cleric Wins Paraguay Presidency, Ending a Party's 62-Year Rule." *New York Times*, April 21, 2008. https://www.nytimes.com/2008/04/21/world/ americas/21paraguay.html.

Barron, Brigid, Caitlin Kennedy Martin, Lori Takeuchi, and Rachel Fithian. "Parents as Learning Partners in the Development of Technological Fluency." *International Journal of Learning and Media* 1, no. 2 (Spring 2009): 55–77. https://doi.org/10.1162/ ijlm.2009.0021.

Barshay, Jill. "Getting boys—and girls—interested in computer coding." *Education By the Numbers* (blog), posted August 12, 2014. http://educationbythenumbers.org/ content/girls-computer-coding_1691/.

Beavis, Catherine, and Claire Charles. "Would the 'Real' Girl Gamer Please Stand Up? Gender, LAN Cafés and the Reformulation of the 'Girl' Gamer." *Gender and Education* 19, no. 6 (2007): 691–705. https://doi.org/10.1080/09540250701650615.

Becker, Howard S. *Art Worlds*. Berkeley: University of California Press, 1982.

Bell, Genevieve, and Paul Dourish. "Yesterday's Tomorrows: Notes on Ubiquitous Computing's Dominant Vision." *Personal and Ubiquitous Computing* 11, no. 2 (February 2007): 133–143. https://doi.org/10.1007/s00779-006-0071-x.

Bender, Walter. "One Laptop per Child: Revolutionizing How the World's Children Engage in Learning." Lecture, MIT Museum, Cambridge, MA, January 17, 2007. Video, 1:11:04, uploaded April 12, 2018. https://techtv.mit.edu/videos/16206- one-laptop-per-child-revolutionizing-how-the-world-s-children-engage-in -learning.

Bender, Walter, Charles Kane, Jody Cornish, and Neal Donahue. *Learning to Change the World: The Social Impact of One Laptop per Child*. New York: Palgrave Macmillan, 2012.

Bennett, Tom. "Sugata Mitra and the Hole in the Research." *TES*, August 1, 2015. https://www.tes.com/news/blog/tom-bennett-sugata-mitra-and-hole-research.

Berger, Bennett M. *The Survival of a Counterculture: Ideological Work and Everyday Life among Rural Communards*. 1981. Reprint, New Brunswick, NJ: Transaction, 2004.

Berlant, Lauren. *Cruel Optimism*. Durham, NC: Duke University Press, 2011.

Bernstein, Robin. *Racial Innocence: Performing American Childhood from Slavery to Civil Rights*. New York: New York University Press, 2011.

Berstein, Brian. "Low-Cost Laptop Program Sees a Key Leadership Defection." *USA Today*, April 22, 2008.

Biba, Erin, and Bobby McKenna. "Turn On, Code In, Drop Out: Tech Programmers Don't Need College Diplomas." *GOOD*, November 19, 2011. https://www.good.is/articles/turn-on-code-in-drop-out.

Bijker, Wiebe E. "Sociohistorical Technology Studies." In *Handbook of Science and Technology Studies*, edited by Sheila Jasanoff, Gerald E. Markle, James C. Petersen, and Trevor Pinch, 229–256. Thousand Oaks, CA: SAGE, 1994.

Bilton, Nick. "Steve Jobs Was a Low-Tech Parent." *New York Times*, September 10, 2014. https://www.nytimes.com/2014/09/11/fashion/steve-jobs-apple-was-a-low-tech-parent.html.

Blikstein, Paulo, chair. Seymour Paper tribute panel at Interaction Design and Children Conference 2013, New School, New York, NY, June 27, 2013. https://tltl.stanford.edu/papert_tribute.

———. "Seymour Papert's Legacy: Thinking About Learning, and Learning About Thinking." Remarks at Seymour Paper tribute panel at Interaction Design and Children Conference 2013, New School, New York, NY, June 27, 2013. https://tltl.stanford.edu/content/seymour-papert-s-legacy-thinking-about-learning-and-learning-about-thinking.

———. "Travels in Troy with Freire: Technology as an Agent of Emancipation." In *Social Justice Education for Teachers: Paulo Freire and the Possible Dream*, edited by Carlos Alberto Torres and Pedro Noguera, 205–244. Rotterdam: Sense, 2008.

———. "You Cannot Think about Thinking without Thinking about What Seymour Papert Would Think." Remarks at Seymour Paper tribute panel at Interaction Design and Children Conference 2013, New School, New York, NY, June 27, 2013. https://tltl.stanford.edu/content/you-cannot-think-about-thinking-without-thinking-about-what-seymour-papert-would-think.

Blikstein, Paulo, and Dor Abrahamson, chairs. "Conference Theme: Logo, the Next 50 Years." IDC 2017 (website). Accessed November 6, 2018. http://idc2017.stanford .edu/conference-theme-logo-the-next-50-years/.

Blizzard, Chris. "Chris Blizzard at OLPC Analyst Meeting." Transcript of remarks at analyst meeting, Cambridge, MA, April 26, 2007. OLPC Talks, 2007. https://web .archive.org/web/20071014182928/http://olpctalks.com/christopher_blizzard/ christopher_blizzard_olpc_meeting.html.

———. "Doom on the XO." *Oxdeadbeef* (blog), November 28, 2006. https://web .archive.org/web/20120507233239/http://www.0xdeadbeef.com/weblog/2006/11/ doom-on-the-xo/

———. "One Laptop per Child and Open Source." *Oxdeadbeef* (blog), May 3, 2007. https://web.archive.org/web/20120508020929/http://www.0xdeadbeef.com/ weblog/2007/05/one-laptop-per-child-and-open-source/

Boulay, Benedict du, and Tim O'Shea. "How to Work the LOGO Machine: A Primer for ELOGO." DAI Occasional Paper No. 4, Department of Artificial Intelligence, University of Edinburgh, Edinburgh, UK, November 1976. http://history.dcs.ed.ac.uk/ archive/docs/how-to-work-the-logo-machine-dai-op-4.pdf.

Bowker, Geoffrey C., Karen Baker, Florence Millerand, and David Ribes. "Toward Information Infrastructure Studies: Ways of Knowing in a Networked Environment." In *International Handbook of Internet Research*, edited by Jeremy Hunsinger, Lisbeth Klastrup, and Matthew Allen, 97–117. Dordrecht, the Netherlands: Springer, 2010. https://doi.org/10.1007/978-1-4020-9789-8_5.

Boym, Svetlana. *The Future of Nostalgia.* New York: Basic Books, 2001.

Brand, Stewart. "Spacewar: Fanatic Life and Symbolic Death among the Computer Bums." *Rolling Stone*, December 1972.

Brand, Stewart. *The Media Lab: Inventing the Future at M.I.T.* New York: Penguin, 1988.

Breitkopf, Antje. "Observations of 'Una Laptop Por Nino' - OLPC Peru." One Laptop per Child, December 15, 2010. http://laptop.org/news/observations-una-laptop-por -nino-olpc-peru.

Brennan, Karen. "Scratch@MIT 2010: Reimagine, Rethink, Remix." ScratchED (forum), 2009–2010. http://scratched.gse.harvard.edu/discussions/events/scratchmit-2010 -reimagine-rethink-remix.

Brown, John Seely, Allan Collins, and Paul Duguid. "Situated Cognition and the Culture of Learning." *Educational Researcher* 18, no. 1 (January 1989): 32–42. https:// doi.org/10.2307/1176008.

Bullis, Kevin. "A Hundred-Dollar Laptop for Hungry Minds." *MIT Technology Review*, September 28, 2005. https://www.technologyreview.com/s/404685/a-hundred-dollar -laptop-for-hungry-minds/.

Burawoy, Michael. "The Extended Case Method." *Sociological Theory* 16, no. 1 (March 1998): 4–33. https://doi.org/10.1111/0735-2751.00040.

Bureau of Democracy Human Rights, and Labor. "2008 Country Reports on Human Rights Practices: Paraguay." US Department of State (website), February 25, 2009. https://www.state.gov/j/drl/rls/hrrpt/2008/wha/119169.htm.

Burrell, Jenna. *Invisible Users: Youth in the Internet Cafés of Urban Ghana*. Cambridge, MA: MIT Press, 2012.

Burrell, Jenna, and Elisa Oreglia. "The Myth of Market Price Information: Mobile Phones and the Application of Economic Knowledge in ICTD." *Economy and Society* 44, no. 2 (2015): 271–292. https://doi.org/10.1080/03085147.2015.1013742.

Butler, Judith. *Excitable Speech: A Politics of the Performative*. New York: Routledge, 1997.

———. *Gender Trouble: Feminism and the Subversion of Identity*. New York: Routledge, 1990.

Callon, Michel. "Some Elements of a Sociology of Translation: Domestication of the Scallops and the Fishermen of St Brieuc Bay." In "Power, Action, and Belief: A New Sociology of Knowledge," edited by John Law. Supplement, *Sociological Review* 32, no. S1 (May 1984): 196–233. https://doi.org/10.1111/j.1467-954X.1984.tb00113.x.

———. "Techno-Economic Networks and Irreversibility." In *A Sociology of Monsters: Essays on Power, Technology, and Domination*, edited by John Law, 132–161. London: Routledge, 1991.

Camfield, Jon. "OLPC History: Senegalese Failure in Implementation." OLPC News, June 8, 2007. http://www.olpcnews.com/people/negroponte/olpc_history_senegal _failure.html.

———. "Total Cost of XO Ownership for OLE Nepal." OLPC News, August 27, 2010. http://www.olpcnews.com/sales_talk/price/total_cost_of_xo_ownership_for.html.

Campbell, Heidi A., and Antonio C. La Pastina. "How the iPhone Became Divine: New Media, Religion and the Intertextual Circulation of Meaning." *New Media & Society* 12, no. 7 (November 2010): 1191–207. https://doi.org/10.1177/1461444810362204.

Cappelle, Frank van, Vanessa Evers, and Sugata Mitra. "Investigating the Effects of Unsupervised Computer Use on Educationally Disadvantaged Children's Knowledge and Understanding of Computers." In *Cultural Attitudes towards Technology and Communication 2004: Proceedings of the Fourth International Conference on Cultural Attitudes towards Technology and Communication, Karlstad, Sweden, 27 June–1 July 2004*, edited by Fay Sudweeks and Charles Ess, 2005, 528–542. Murdoch, Australia: School of Information Technology, Murdoch University, 2004.

Ceaser, Mike. "Paraguay's Archive of Terror." *BBC News*, March 11, 2002. http://news
.bbc.co.uk/2/hi/americas/1866517.stm.

Cellan-Jones, Rory. "Negroponte—Throwing the Laptop." YouTube video, 0:41, published December 18, 2008. https://www.youtube.com/watch?v=-tqFEzDou6s.

Central Intelligence Agency. "Paraguay." *The World Factbook*. 2012. https://www.cia
.gov/library/publications/the-world-factbook/geos/pa.html

Ceruzzi, Paul E. "Moore's Law and Technological Determinism: Reflections on the History of Technology." *Technology and Culture* 46, no. 3 (July 2005): 584–593. https://doi.org/10.1353/tech.2005.0116.

Chan, Anita Say. "Coding Free Software, Coding Free States: Free Software Legislation and the Politics of Code in Peru." *Anthropological Quarterly* 77, no. 3 (Summer 2004): 531–545. https://doi.org/10.1353/anq.2004.0046.

———. "Hacking Digital Universalism: Reconfiguring Design at the Periphery." In *Beyond Imported Magic: Science and Technology Studies in Latin America*, edited by Ivan da Costa Marques, Christina Holmes, and Eden Medina. Cambridge, MA: MIT Press, 2014.

———. *Networking Peripheries: Technological Futures and the Myth of Digital Universalism*. Cambridge, MA: MIT Press, 2014.

Christensen, Gayle, Andrew Steinmetz, Brandon Alcorn, Amy Bennett, Deirdre Woods, and Ezekiel J. Emanuel. "The MOOC Phenomenon: Who Takes Massive Open Online Courses and Why?" Working paper, November 6, 2013. https://doi
.org/10.2139/ssrn.2350964.

Chudacoff, Howard P. *Children at Play: An American History*. New York; New York University Press, 2007.

Citron, Danielle Keats. *Hate Crimes in Cyberspace*. Cambridge, MA: Harvard University Press, 2014.

Clammer, John. "Performing Development: Theatre of the Oppressed and Beyond." In *Art, Culture and International Development: Humanizing Social Transformation*, 63–82. Andover, UK: Routledge, 2014.

Clark, Donald. "More Holes in Sugata Mitra's 'Hole-in-Wall' Project." *Donald Clark Plan B* (blog), June 2, 2013. http://donaldclarkplanb.blogspot.com/2013/06/more
-holes-in-sugata-mitras-hole-in.html.

———. "Sugata Mitra: Slum Chic? 7 Reasons for Doubt." *Donald Clark Plan B* (blog), March 4, 2013. http://donaldclarkplanb.blogspot.com/2013/03/sugata-mitra-slum
-chic-7-reasons-for.html.

Clements, Douglas. "Research on Logo in Education: Is the Turtle Slow but Steady, or Not Even in the Race?" *Computers in the Schools* 2, no. 2–3 (1985): 55–71. https://doi.org/10.1300/J025v02n02_07.

Clements, Douglas H., and Dominic F. Gullo. "Effects of Computer Programming on Young Children's Cognition." *Journal of Educational Psychology* 76, no. 6 (December 1984): 1051–1058. https://doi.org/10.1037/0022-0663.76.6.1051.

Clements, Douglas H., and Julie Sarama. "Research on Logo: A Decade of Progress." *Computers in the Schools* 14, no. 1–2 (1997): 9–46. https://doi.org/10.1300/J025v14n01_02.

Clow, Doug. "MOOCs and the Funnel of Participation." In *Proceedings of the Third International Conference on Learning Analytics and Knowledge*, edited by Dan Suthers, Katrien Verbert, Erik Duval, and Xavier Ochoa, 185–189. New York: ACM, 2013. https://doi.org/10.1145/2460296.2460332.

Coleman, E. Gabriella. *Coding Freedom: The Ethics and Aesthetics of Hacking*. Princeton, NJ: Princeton University Press, 2013.

———. "The Hacker Conference: A Ritual Condensation and Celebration of a Lifeworld." *Anthropological Quarterly* 83, no. 1 (Winter 2010): 47–72. https://doi.org/10.1353/anq.0.0112.

Coleman, E. Gabriella, and Alex Golub. "Hacker Practice: Moral Genres and the Cultural Articulation of Liberarlism." *Anthropological Theory* 8, no. 3 (September 2008): 255–277. https://doi.org/10.1177/1463499608093814.

Condliffe, Jamie. "Ethiopian Kids Hack Their OLPC in 5 Months, with No Help." *Gizmodo*, October 31, 2012. https://gizmodo.com/5956417/ethiopian-kids-hack-their-olpc-tablets-in-5-months-with-no-help.

Coontz, Stephanie. *The Way We Never Were: American Families and the Nostalgia Trap*. New York: Basic Books, 1992.

Cowan, Ruth Schwartz. "The 'Industrial Revolution' in the Home: Household Technology and Social Change in the 20th Century." *Technology and Culture* 17, no. 1 (January 1976): 1–23. https://doi.org/10.2307/3103251.

Cristia, Julián P., Pablo Ibarrarán, Santiago Cueto, Ana Santiago, and Eugenio Severín. "Technology and Child Development: Evidence from the One Laptop per Child Program." Discussion Paper No. 6401, Institute for the Study of Labor, Bonn, Germany, March 2012. http://repec.iza.org/dp6401.pdf.

Cuban, Larry. *The Blackboard and the Bottom Line: Why Schools Can't Be Businesses*. Cambridge, MA: Harvard University Press, 2004.

———. "No End to Magical Thinking When It Comes to High-Tech Schooling." *Larry Cuban on School Reform and Classroom Practice* (blog), March 18, 2013.

https://larrycuban.wordpress.com/2013/03/18/no-end-to-magical-thinking-when
-it-comes-to-high-tech-schooling/.

———. *Oversold and Underused: Computers in the Classroom.* Cambridge, MA: Harvard
University Press, 2001.

———. "Reflections on 'How to Break Free from Our 19th Century Factory-Model
Education.'" *Larry Cuban on School Reform and Classroom Practice* (blog), July 19, 2012.
https://larrycuban.wordpress.com/2012/07/19/reflections-on-how-to-break-free
-from-our-19th-century-factory-model-education/.

———. *Teachers and Machines: The Classroom Use of Technology since 1920.* New York:
Teachers College Press, 1986.

Dafoe, Allan. "On Technological Determinism: A Typology, Scope Conditions, and
a Mechanism." *Science, Technology & Human Values* 40, no. 6 (2015): 1047–1076.
https://doi.org/10.1177/0162243915579283.

Derndorfer, Christoph. "An Interview with Sandro Marcone about Peru's Una
Laptop por Niño." OLPC News, October 8, 2012. http://www.olpcnews.com/
countries/peru/english_summary_at_the_end.html.

———. "OLPC in Paraguay: Will ParaguayEduca's XO Laptop Deployment Success
Scale?" *Education Technology Debate* (blog), October 19, 2010. http://edutechdebate
.org/olpc-in-south-america/will-paraguayeduca-scale/.

———. "OLPC in Peru: A Problematic Una Laptop por Niño Program." *Education
Technology Debate* (blog), October 27, 2010. http://edutechdebate.org/olpc-in-south
-america/olpc-in-peru-one-laptop-per-child-problems/.

———. "Plan Ceibal Expands New Repair System to Address High XO Breakage
Rates." OLPC News, December 7, 2011. http://www.olpcnews.com/countries/
uruguay/plan_ceibal_expands_new_repair_system_to_address_high_breakage_rates
.html.

———. "Plan Ceibal's 4-Year Cost Increases from $276 to $400." OLPC News, January
10, 2012. http://www.olpcnews.com/countries/uruguay/plan_ceibals_4_year_cost
_increases_from_276_to_400.html.

———. "Where Have All the OLPC School Servers Gone?" OLPC News, May 29,
2007. http://www.olpcnews.com/hardware/school_servers/olpc_school_servers
.html.

Dillahunt, Tawanna, Zengguang Wang, and Stephanie D. Teasley. "Democratizing Higher Education: Exploring MOOC Use among Those Who Cannot Afford a
Formal Education." *International Review of Research in Open and Distance Learning* 15,
no. 5 (2014): 177–196. https://doi.org/10.19173/irrodl.v15i5.1841.

Dobrin, Sidney I. *Constructing Knowledges: The Politics of Theory-Building and Pedagogy in Composition*. Albany: State University of New York Press, 1997.

Douglas, Susan J. *Listening In: Radio and the American Imagination*. 1999. Reprintg, Minneapolis: University of Minnesota Press, 2004.

Dourish, Paul, and Genevieve Bell. *Divining a Digital Future: Mess and Mythology in Ubiquitous Computing*. Cambridge, MA: MIT Press, 2011.

———. "'Resistance Is Futile': Reading Science Fiction alongside Ubiquitous Computing." *Personal and Ubiquitous Computing* 18, no. 4 (April 2014): 769–778. https://doi.org/10.1007/s00779-013-0678-7.

Drake, Daniel. Posts tagged "Paraguay." *dsd's weblog*. Accessed November 6, 2018. http://www.reactivated.net/weblog/archives/tag/paraguay/.

———. "XO Laptop Deployment Logistics." *dsd's weblog*, May 2, 2009. http://www.reactivated.net/weblog/archives/2009/05/xo-laptop-deployment-logistics/.

Dray, James, and Joseph A. Menosky. "Computers and a New World Order." *MIT Technology Review*, May/June 1983, 12–16.

Dunckley, Victoria. L. "Gray Matters: Too Much Screen Time Damages the Brain." *Mental Wealth* (blog). *Psychology Today*, February 27, 2014. https://www.psychologytoday.com/us/blog/mental-wealth/201402/gray-matters-too-much-screen-time-damages-the-brain.

Dunne, Anthony, and Fiona Raby. *Speculative Everything: Design, Fiction, and Social Dreaming*. Cambridge, MA: MIT Press, 2013.

Dupuy, Jean-Pierre. *On the Origins of Cognitive Science: The Mechanization of the Mind*. Cambridge, MA: MIT Press, 2009.

Echikson, William. "Microcomputer Center in Paris." *Christian Science Monitor*, June 2, 1982. https://www.csmonitor.com/1982/0602/060224.html.

Economist. "The Boy and the Bishop: Paternity Claims Distract from a Struggle for Reform." April 30, 2009. https://www.economist.com/the-americas/2009/04/30/the-boy-and-the-bishop.

———. "Error Message: A Disappointing Return from an Investment in Computing." April 7, 2012. https://www.economist.com/the-americas/2012/04/07/error-message.

———. "The Next Leftist on the Block: Measuring up Fernando Lugo's Plans for a Misgoverned Country." August 7, 2008. https://www.economist.com/the-americas/2008/08/07/the-next-leftist-on-the-block.

Edwards, Paul N. "The Closed World: Systems Discourse, Military Policy and Post–World War II US Historical Consciousness." In *Cyborg Worlds: The Military*

Information Society, edited by Les Levidow and Kevin Robins, 135–158. London: Free Association Books, 1989.

Edwards, Stephen A., and Martha A. Kim. "History of Processor Performance." Lecture notes for April 23, 2012, CSEE W3827: Fundamentals of Computer Systems, Spring 2012, Columbia University, New York, NY. http://www.cs.columbia.edu/~sedwards/classes/2012/3827-spring/advanced-arch-2011.pdf.

El Observador (Uruguay). "Se entregó la ceibalita un millón." October 2, 2013. https://www.elobservador.com.uy/nota/-se-entrego-la-ceibalita-un-millon-201310217320.

Electronic Frontier Foundation. "Content Blocking." Accessed November 6, 2018. https://www.eff.org/issues/content-blocking.

Engle, Deborah, Chris Mankoff, and Jennifer Carbrey. "Coursera's Introductory Human Physiology Course: Factors That Characterie Successful Completion of a MOOC." *International Review of Research in Open and Distributed Learning* 16, no. 2 (April 2015): 46–68. http://dx.doi.org/10.19173/irrodl.v16i2.2010.

Ensmenger, Nathan L. "'Beards, Sandals, and Other Signs of Rugged Individualism': Masculine Culture within the Computing Professions." *Osiris* 30 (2015): 38–65. https://doi.org/10.1086/682955.

———. *The Computer Boys Take Over: Computers, Programmers, and the Politics of Technical Expertise*. Cambridge, MA: MIT Press, 2012.

———. "Letting the 'Computer Boys' Take Over: Technology and the Politics of Organizational Transformation." In "Uncovering Labour in Information Revolutions, 1750–2000," edited by Aad Blok and Greg Downey. Supplement, *International Review of Social History* 48, no. S11 (2003): 153–180. https://doi.org/10.1017/S0020859003001305.

———. "Making Programming Masculine." *Gender Codes: Why Women Are Leaving Computing*, edited by Thomas J. Misa, 115–141. Hoboken, NJ: John Wiley & Sons, 2010. https://doi.org/10.1002/9780470619926.ch6.

Fail Festival. "Why." Accessed November 6, 2018. http://failfestival.org/why/.

Farivar, Cyrus. "Waiting for That $100 Laptop? Don't Hold Your Breath." *Slate*, November 29, 2005. https://slate.com/culture/2005/11/the-mythical-100-laptop.html.

Federal Research Division, Library of Congress. "Country Profile: Paraguay." Library of Congress (website), October 2005. https://www.loc.gov/rr/frd/cs/profiles/Paraguay.pdf.

Fejes, Fred. "Media Imperialism: An Assessment." *Media, Culture & Society* 3, no. 3 (July 1981): 281–289. https://doi.org/10.1177/016344378100300306.

Ferguson, James. *Global Shadows: Africa in the Neoliberal World Order*. Durham, NC: Duke University Press, 2006.

Finkle, Caryn L. "Nestlé, Infant Formula, and Excuses: The Regulation of Commercial Advertising in Developing Nations." *Northwestern Journal of International Law & Business* 14, no. 3 (Spring 1994): 602–619. https://scholarlycommons.law .northwestern.edu/njilb/vol14/iss3/31.

Fisher, Ken. "OLPC's XO Laptop Comes with Anti-Theft Kill-Switch in Select Countries." *Ars Technica*, February 18, 2007. https://arstechnica.com/uncategorized/ 2007/02/8872/.

Fortunati, Leopoldina. "The Mobile Phone: An Identity on the Move." *Personal and Ubiquitous Computing* 5, no. 2 (July 2001): 85–98. https://doi.org/10.1007/ PL00000017.

Fouché, Rayvon. "From Black Inventors to One Laptop per Child: Exporting a Racial Politics of Technology." In *Race after the Internet*, edited by Lisa Nakamura and Peter Chow-White, 61–83. New York: Routledge, 2011.

———. "Say It Loud, I'm Black and I'm Proud: African Americans, American Artifactual Culture, and Black Vernacular Technological Creativity." *American Quarterly* 58, no. 3 (September 2006): 639–661. https://doi.org/10.1353/aq.2006.0059.

Freire, Paulo. *Pedagogy of the Oppressed*. Translated by Myra Bergman Ramos. New York: Herder and Herder, 1970.

Fried, Ina. "Negroponte: You Really Can Give a Kid a Laptop." *CNet*, August 6, 2010. https://www.cnet.com/news/negroponte-you-really-can-give-a-kid-a-laptop/.

Gacitúa Marió, Estanislao, Annika Silva-Leander, and Miguel Carter. *Paraguay: Social Development Issues for Poverty Alleviation: Country Social Analysis*. Social Development Papers No. 63. Washington, DC: World Bank, 2004. http://documents .worldbank.org/curated/en/448491468763205050/Paraguay-social-development -issues-for-poverty-alleviation-country-social-analysis.

Gay, Paul du, Stuart Hall, Linday Janes, Hugh Mackay, and Keith Negus. *Doing Cultural Studies: The Story of the Sony Walkman*. London: SAGE, 1997.

Geertz, Clifford. "Thick Description: Towards an Interpretive Theory of Culture." In *The Interpretation of Cultures: Selected Essays*, 1–30. New York: Basic Books, 1973.

GettyImages. "UN World Summit on the Information Society Photos: Negroponte," 2005. https://www.gettyimages.ae/photos/world-summit-on-the-information-society -2005-negroponte/.

Gibson, William. *Neuromancer*. New York: Ace Books, 1984.

Goffman, Erving. *The Presentation of Self in Everyday Life*. Harmondsworth, UK: Penguin, 1959.

Gonzalez, Laura. "When Two Tribes Go to War: A History of Video Game Controversy." *Gamespot*, January 10, 2007. https://www.gamespot.com/articles/when-two-tribes-go -to-war-a-history-of-video-game-controversy/1100-6090892/.

Gray, Kishonna L. "Intersecting Oppressions and Online Communities: Examining the Experiences of Women of Color in Xbox Live." *Information, Communication & Society* 15, no. 3 (2012): 411–428. https://doi.org/10.1080/1369118X.2011.642401.

Guernsey, Lisa. *Into the Minds of Babes: How Screen Time Affects Children from Birth to Age Five*. New York: Basic Books, 2007.

Gugler, Josef. "Fiction, Fact, and the Critic's Responsibility: *Camp de Thiaroye, Yaaba*, and *The Gods Must Be Crazy*." In *Focus on African Films*, edited by Françoise Pfaff, 69–85. Bloomington: Indiana University Press, 2004.

Gutman, Roy. *Banana Diplomacy: The Making of American Policy in Nicaragua, 1981– 1987*. New York: Simon and Schuster, 1988.

Hachman, Mark. "Negroponte: We'll Throw OLPCs out of Helicopters to Teach Kids to Read." *PC Magazine*, November 2, 2011. https://www.pcmag.com/article2/ 0,2817,2395763,00.asp.

Hager, Chris. "[Olpcaustria] xo-get & svg-grabber." OLPC Austria mailing list, December 6, 2007. http://lists.lo-res.org/pipermail/olpcaustria/2007-December/000453 .html.

Hall, Stuart. "Encoding/Decoding." In *The Cultural Studies Reader*, edited by Simon During, 90–103. London: Routledge, 1993.

———. "The Rediscovery of 'Ideology': The Return of the Repressed in Media Studies." In *Culture, Society, and the Media*, edited by Michael Gurevitch, Tony Bennett, James Curran, and Janet Woollacott, 52–86. London: Methuen, 1982.

Haraway, Donna J. *Staying with the Trouble: Making Kin in the Chthulucene*. Durham, NC: Duke University Press, 2016.

Haring, Kristen. *Ham Radio's Technical Culture*. Cambridge, MA: MIT Press, 2007.

Harvey, David. "The Fetish of Technology: Causes and Consequences." *Macalester International* 13 (2003): 3–30. https://digitalcommons.macalester.edu/macintl/vol13/ iss1/7/.

Hassan, Robert. "The MIT Media Lab: Techno Dream Factory or Alienation as a Way of Life?" *Media, Culture & Society* 25, no. 1 (January 2003): 87–106. https://doi .org/10.1177/016344370302500106.

Hempel, Carlene. "The Miseducation of Young Nerds." *Greensboro News & Record*, February 3, 2000. https://www.greensboro.com/the-miseducation-of-young-nerds -given-a-choice-technologically-inclined/article_cd46f3b8-0f14-5984-a506 -d158182b2738.html

Hicks, Marie. "De-brogramming the History of Computing." *IEEE Annals of the History of Computing* 35, no. 1 (January–March 2013): 86–88. https://doi.org/10.1109/ MAHC.2013.3.

————. *Programmed Inequality: How Britain Discarded Women Technologists and Lost Its Edge in Computing*. Cambridge, MA: MIT Press, 2017.

Hill, Benjamin Mako. "The Geek Shall Inherit the Earth: My Story of Unlearning." Benjamin Mako Hill (personal website), November 18, 2002. Last modified March 15, 2013. https://mako.cc/writing/unlearningstory/StoryOfUnlearing.html.

————. "Laptop Liberation." *Copyrighteous* (blog), April 29, 2008. http://mako.cc/ copyrighteous/laptop-liberation-2.

————. "OLPC and Charges of Technological and Cultural Imperialism." *Copyrighteous* (blog), December 15, 2005. https://mako.cc/copyrighteous/olpc-and-charges -of-technological-and-cultural-imperialism.

Himanen, Pekka. *The Hacker Ethic: A Radical Approach to the Philosophy of Business*. New York: Random House, 2001.

Hodzic, Saida. "Performing Development: Women's NGOs, Donors, and the Postcolonial Ghanaian State." PhD diss., University of California, San Francisco with the University of California, Berkeley, 2006.

Hornborg, Alf. "Technology as Fetish: Marx, Latour, and the Cultural Foundations of Capitalism." *Theory, Culture & Society* 31, no. 4 (July 2014): 119–140. https://doi .org/10.1177/0263276413488960.

Horowitz, Roger, ed. *Boys and Their Toys? Masculinity, Technology, and Class in America*. New York: Routledge, 2001.

Humphreys, Lee. "Technological Determinism." In *Encyclopedia of Science and Technology Communication*, edited by Susanna Hornig Priest, 869–872. Los Angeles: SAGE, 2012.

Innocenti, Bernie. "Brain Dump." Codewiz (personal website). Accessed November 6, 2018. http://codewiz.org/wiki/BrainDump.

Irani, Lilly. "Hackathons and the Making of Entrepreneurial Citizenship." *Science, Technology & Human Values* 40, no. 5 (September 2015): 799–824. https://doi.org/10 .1177/0162243915578486.

Ito, Mizuko. *Engineering Play: A Cultural History of Children's Software*. Cambridge, MA: MIT Press, 2009.

Ito, Mizuko, Sonja Baumer, Matteo Bittanti, danah boyd, Rachel Cody, Becky Herr-Stephenson, Heather A. Horst, et al. *Hanging Out, Messing Around, and Geeking Out: Kids Living and Learning with New Media*. Cambridge, MA: MIT Press, 2010.

Ito, Mizuko, Kris Gutiérrez, Sonia Livingstone, Bill Penuel, Jean Rhodes, Katie Salor, Juliet Schor, Julian Sefton-Green, and S. Craig Watkins. *Connected Learning: An Agenda for Research and Design*. Irvine, CA: Digital Media and Learning Research Hub, 2013. https://dmlhub.net/publications/connected-learning-agenda-for-research-and-design/.

Ito, Mizuko, Heather Horst, Matteo Bittanti, danah boyd, Becky Herr-Stephenson, Patricia G. Lange, C. J. Pascoe, et al. *Living and Learning with New Media: Summary of Findings from the Digital Youth Project*. Cambridge, MA: MIT Press, 2008.

Jasanoff, Sheila, and San-Hyun Kim, eds. *Dreamscapes of Modernity: Sociotechnical Imaginaries and the Fabrication of Power*. Chicago: University of Chicago Press, 2015.

Jasanoff, Sheila, Sang-Hyun Kim, and Stefan Sperling. "Sociotechnical Imaginaries and Science and Technology Policy: A Cross-National Comparison." Research proposal, September 2007. https://www.researchgate.net/publication/265664653.

Jenkins, Henry, with Ravi Purushotma, Margaret Weigel, Katie Clinton, and Alice J. Robison. *Confronting the Challenges of Participatory Culture: Media Education for the 21st Century*. Cambridge, MA: MIT Press, 2006.

Johnson, Jim. "Mixing Humans and Nonhumans Together: The Sociology of a Door-Closer." *Social Problems* 35, no. 3 (June 1988): 298–310. https://doi.org/10.1525/sp.1988.35.3.03a00070.

Kaplan, David A. "Sal Khan: Bill Gates' Favorite Teacher." *CNN Money*, August 24, 2010. https://money.cnn.com/2010/08/23/technology/sal_khan_academy.fortune/index.htm.

Kasumovic, Michael M., and Jeffrey H. Kuznekoff. "Insights into Sexism: Male Status and Performance Moderates Female-Directed Hostile and Amicable Behaviour." *PLoS ONE* 10, no. 7 (2015): e0131613. https://doi.org/10.1371/journal.pone.0131613.

Keller, Matt, Edward McNierney, and Richard Smith. "The Reading Project." Presentation at the OLPC San Francisco Community Summit 2012, San Francisco, CA, October 21, 2012. https://www.olpcsf.org/CommunitySummit2012/sessions/the-reading-project.

Kelty, Christopher. "Geeks, Social Imaginaries, and Recursive Publics." *Cultural Anthropology* 20, no. 2 (May 2005): 185–214. https://doi.org/10.1525/can.2005.20.2.185.

Kessen, William. "The American Child and Other Cultural Inventions." *American Psychologist*, 34, no. 10 (October 1979): 815–820. https://doi.org/10.1037/0003-066X.34.10.815.

Khalil, Hanan, and Martin Ebner. "MOOCs Completion Rates and Possible Methods to Improve Retention: A Literature Review." *Proceedings of EdMedia + Innovate Learning 2014*, edited by Jan Herrington, Jarmo Viteli and Marianna Leikomaa, 1305–1313. Waynesville, NC: Association for the Advancement of Computing in Education, 2014.

Khan, Sal. "Let's Use Video to Reinvent Education." Filmed March 2011 in Long Beach, CA. TED video, 20:27. https://www.ted.com/talks/salman_khan_let_s_use_video_to_reinvent_education.

Kidwell, Peggy Aldrich. "Stalking the Elusive Computer Bug." *IEEE Annals of the History of Computing* 20, no. 4 (October–December 1998): 5–9. https://doi.org/10.1109/85.728224.

King, Brad, and John Borland. *Dungeons and Dreamers: The Rise of Computer Game Culture from Geek to Chic*. Emeryville, CA: McGraw-Hill/Osborne, 2003.

Kirkpatrick, David. "World's First Working $100 Laptop." *CNN Money*, November 16, 2005. https://money.cnn.com/2005/11/16/technology/laptop_fortune/.

Kirkpatrick, Graeme. "How Gaming Became Sexist: A Study of UK Gaming Magazines, 1981–1995." *Media, Culture & Society* 39, no. 4 (May 2017): 453–468. https://doi.org/10.1177/0163443716646177.

Kirsch, Scott. "The Incredible Shrinking World? Technology and the Production of Space." *Environment and Planning D: Society and Space* 13, no. 5 (1995): 529–555. https://doi.org/10.1068/d130529.

Kleiman, Michael, dir. *Web*. 2013. Documentary Movie, 1:22:32, YouTube streaming. https://www.youtube.com/watch?v=wMzMFuxpqJU

Kling, Rob, ed. *Computerization and Controversy: Value Conflicts and Social Choices*. 2nd ed. San Diego: Morgan Kaufmann, 1996.

Knight, Will. "$100-Laptop Created for World's Poorest Countries." *New Scientist*, November 17, 2005. https://www.newscientist.com/article/dn8338-100-laptop-created-for-worlds-poorest-countries/.

Koschmann, Timothy. "Logo-as-Latin Redux." Review of *The Children's Machine: Rethinking School in the Age of the Computer*, by Seymour Papert. *Journal of the Learning Sciences* 6, no. 4 (1997): 409–415. https://www.jstor.org/stable/1466780.

Krstić, Ivan. "Google EngEDU Tech Talk: Ivan Krstić," April 12, 2007. https://web.archive.org/web/20110228165736/http://www.olpctalks.com:80/ivan_krsti/

ivan_krstic_at_google.html   and   https://web.archive.org/web/20071017070317/
http://www.olpctalks.com/ivan_krsti/ivan_krstic_talks.html

————. "Maintaining Clarity." Ivan Krstić (personal website), March 18, 2008.
https://web.archive.org/web/20080320203548/http://radian.org:80/notebook/
maintaining-clarity

————. "Sic Transit Gloria Laptopi." Ivan Krstić (personal website), May 13, 2008.
https://web.archive.org/web/20080517102347/http://radian.org:80/notebook/
sic-transit-gloria-laptopi

————. "Sweet Nonsense Omelet." Ivan Krstić (personal website), July 22, 2009.
https://web.archive.org/web/20090726005526/http://radian.org:80/notebook/
nonsense-omelet

————. "This, Too, Shall Pass, or: Things to Remember When Reading News about
OLPC." Ivan Krstić (personal website), April 25, 2008. https://web.archive.org/
web/20140420030324/http://radian.org/notebook/this-too-shall-pass

Kurland, D. Midian, and Roy D. Pea. "Children's Mental Models of Recursive Logo
Programs." *Journal of Educational Computing Research* 1, no. 2 (May 1985): 235–243.
https://doi.org/10.2190/JV9Y-5PD0-MX22-9J4Y.

La Nación (Paraguay). "Alumna del cuarto grado ganó concurso mundial en Nickelodeon." September 27, 2011.

Landa, Robin. *Graphic Design Solutions.* 5th ed. Boston: Wadsworth, 2014.

Lareau, Annette. "Social Class Differences in Family-School Relationships: The
Importance of Cultural Capital." *Sociology of Education* 60, no. 2 (April 1987): 73–85.
http://doi.org/10.2307/2112583.

Lareau, Annette, and Wesley Shumar. "The Problem of Individualism in Family-
School Policies." In "Special Issue on Sociology and Educational Policy: Bringing
Scholarship and Practice Together," edited by Peter W. Cookson Jr., Joseph C.
Conaty, and Harold S. Himmelfarb. Extra issue, *Sociology of Education* 69 (1996): 24–
39. http://doi.org/10.2307/3108454.

Larkin, Brian. "The Politics and Poetics of Infrastructure." *Annual Review of Anthropology* 42 (2013): 327–343. https://doi.org/10.1146/annurev-anthro-092412-155522.

Latour, Bruno. *Reassembling the Social: An Introduction to Actor-Network-Theory.*
Oxford: Oxford University Press, 2005.

Lauermann, John. "Performing Development in Street Markets: Hegemony, Govern-
mentality, and the Qat Industry of Sana'a, Yemen." *Antipode* 44, no. 4 (September
2012): 1329–1347. https://doi.org/10.1111/j.1467-8330.2011.00955.x.

Lave, Jean, and Étienne Wenger. *Situated Learning: Legitimate Peripheral Participation.* Cambridge: Cambridge University Press, 1991.

Lave, Jean, and Ray McDermott. "Estranged ~~Labor~~ Learning." *Outlines* 4, no. 1 (2002): 19–48. https://tidsskrift.dk/outlines/article/view/5143/4543.

Law, John. "Introduction: Monsters, Machines, and Sociotechnical Relations." In *A Sociology of Monsters: Essays on Power, Technology, and Domination*, edited by John Law, 1–23. London: Routledge, 1991.

Lawler, Robert W., and Masoud Yazdani, eds. *Artificial Intelligence and Education.* Vol. 1, *Learning Environments and Tutoring Systems.* Norwood, NJ: Ablex, 1987.

Lazar, Allan, Dan Karlan, and Jeremy Salter. *The 101 Most Influential People Who Never Lived.* New York: HarperCollins, 2006.

Lee, Richard. "The Gods Must Be Crazy, but the State Has a Plan: Government Policies towards the San in Namibia." *Canadian Journal of African Studies* 20, no. 1 (1986): 91–98. https://doi.org/10.2307/484697.

Lego. "History - Mindstorms." Accessed November 6, 2018. https://www.lego.com/en-us/mindstorms/history.

Levy, Steven. *Hackers: Heroes of the Computer Revolution.* Garden City, NY: Anchor, 1984.

Lienhard, John H. *Inventing Modern: Growing Up with X-rays, Skyscrapers, and Tailfins.* Oxford: Oxford University Press, 2005.

Lisberger, Steven, dir. *Tron.* Buena Vista, 1982.

Liyanagunawardena, Tharindu Rekha, Andrew Alexandar Adams, and Shirley Ann Williams. "MOOCs: A Systematic Study of the Published Literature, 2008–2012." *International Review of Research in Open and Distance Learning* 14, no. 3 (July 2013): 202–227. https://doi.org/10.19173/irrodl.v14i3.1455.

Locke, John. *Some Thoughts Concerning Education.* London, 1693.

Lowes, Susan, and Cyrus Luhr. *Evaluation of the Teaching Matters One Laptop per Child (XO) Pilot at Kappa IV.* Institute for Learning Technologies, Teachers College, Columbia University, June 2008. http://www.teachingmatters.org/files/olpc_kappa.pdf.

Luyt, Brendan. "The One Laptop per Child Project and the Negotiation of Technological Meaning." *First Monday* 13, no. 6 (June 2, 2008). https://doi.org/10.5210/fm.v13i6.2144.

Lydon, Christopher. "One Laptop per Child?" *Radio Open Source*, February 20, 2007. http://radioopensource.org/one-laptop-per-child/.

MacKenzie, Donald. *An Engine, Not a Camera: How Financial Models Shape Markets.* Cambridge, MA: MIT Press, 2006.

Malamud, Ofer, and Cristian Pop-Eleches. "Home Computer Use and the Development of Human Capital." *Quarterly Journal of Economics* 126, no. 2 (May 2011): 987–1027. https://doi.org/10.1093/qje/qjr008.

Mansell, Robin. *Imagining the Internet: Communication, Innovation, and Governance.* Oxford: Oxford University Press, 2012.

Margolis, Jane, and Allan Fisher. *Unlocking the Clubhouse: Women in Computing.* Cambridge, MA: MIT Press, 2003.

Markoff, John. "Taking the Pulse of Technology at Davos." *New York Times,* January 31, 2005. https://www.nytimes.com/2005/01/31/technology/taking-the-pulse-of-technology-at-davos.html.

Martínez, Ana Laura, Serrana Alonso, and Diego Díaz. "Monitoreo y evaluación de impacto social del Plan CEIBAL: Metodología y primeros resultados a nivel nacional." PowerPoint presentation, 2009. https://docplayer.es/40661752-Monitoreo-y-evaluacion-de-impacto-social-del-plan-ceibal.html.

Martinez-Gomez, David, Jared Tucker, Kate A. Heelan, Gregory J. Welk, and Joey C. Eisenmann. "Associations between Sedentary Behavior and Blood Pressure in Young Children." *Archives of Pediatrics and Adolescent Medicine* 163, no. 8 (August 3, 2009): 724–730. https://doi.org/10.1001/archpediatrics.2009.90.

Marx, Karl, and Frederick Engels. *The German Ideology: Part One, with Selections from Parts Two and Three, Together with Marx's "Introduction to a Critique of Political Economy."* Edited and with an introduction by C. J. Arthur. New York: International Publishers, 1970.

Mazzarella, William. "Beautiful Balloon: The Digital Divide and the Charisma of New Media in India." *American Ethnologist* 37, no. 4 (November 2010): 783–804. https://doi.org/10.1111/j.1548-1425.2010.01285.x.

McIntosh, Donald. "Weber and Freud: On the Nature and Sources of Authority." *American Sociological Review* 35, no. 5 (October 1970): 901–911. https://doi.org/10.2307/2093300.

McLuhan, Marshall. *Understanding Media: The Extensions of Man.* 1964. Reprint, Cambridge, MA: MIT Press, 1994.

Medina, Eden. *Cybernetic Revolutionaries: Technology and Politics in Allende's Chile.* Cambridge, MA: MIT Press, 2011.

Medina, Eden, Ivan da Costa Marques, and Christina Holmes, eds. *Beyond Imported Magic: Essays on Science, Technology, and Society in Latin America.* Cambridge, MA: MIT Press, 2014.

Melo, Gioia de, Alina Machado, Alfonso Miranda, and Magdalena Viera. "Profundizando en los efectos del Plan Ceibal." Paper presented at IV Jornadas Académicas, XV Jornadas de Coyuntura Económica de la Facultad de Ciencias Económicas y de Administración, Universidad de la República, Montevideo, Uruguay, August 29, 2013. http://fcea.edu.uy/Jornadas_Academicas/2013/file/MESAS/Economia%20 de%20la%20educacion_plan%20ceibal/Profundizando%20en%20los%20 efectos%20del%20Plan%20Ceibal.pdf.

Mentor, The [Loyd Blankenship]. "The Conscience of a Hacker." January 8, 1986. *Phrack*, no. 7 (September 25, 1986). http://www.phrack.org/issues/7/3.html.

Mills, C. Wright. 1959. *The Sociological Imagination*. New York: Oxford University Press, 40th anniversary edition, 2000.

Mintz, Steven. *Huck's Raft: A History of American Childhood*. Cambridge, MA: Belknap Press of Harvard University Press, 2004.

MIT News. "Annan Presents Prototype $100 Laptop at World Summit on Information Society." November 16, 2005. http://news.mit.edu/2005/laptop-1116.

Mitra, Sugata. *Beyond the Hole in the Wall: Discover the Power of Self-Organized Learning*. Foreword by Nicholas Negroponte. New York: TED Conferences, 2012.

———. "Minimally Invasive Education: A Progress Report on the 'Hole-in-the-Wall' Experiments." *British Journal of Educational Technology* 34, no. 3 (June 2003): 367–371. https://doi.org/10.1111/1467-8535.00333.

Mitra, Sugata, Ritu Dangwa, Shiffon Chatterjee, Swati Jha, Ravinder S. Bisht, and Preeti Kapur. "Acquisition of Computing Literacy on Shared Public Computers: Children and the 'Hole in the Wall.'" *Australasian Journal of Educational Technology* 21, no. 3 (2005): 407–426. https://doi.org/10.14742/ajet.1328.

Montevideo.com. "Sobre el cielo austral: Ciudadanía digital." *Red Especial Uruguaya* (blog), November 30, 2010. http://redespecialuruguaya.blogspot.com/2011/02/ sobre-el-cielo-austral-ciudadania.html.

Mosco, Vincent. *The Digital Sublime: Myth, Power, and Cyberspace*. Cambridge, MA: MIT Press, 2005.

Murph, Darren. "OLPC XO Caught Playing Super Mario Bros. 3." *Engadget*, December 25, 2006. https://www.engadget.com/2006/12/25/olpc-xo-caught-playing-super -mario-bros-3/.

Nagle, Angela. *Kill All Normies: Online Culture Wars from 4Chan and Tumblr to Trump and the Alt-Right*. Winchester, UK: Zero Books, 2017.

Nakamura, Lisa. *Cybertypes: Race, Ethnicity, and Identity on the Internet*. New York: Routledge, 2002.

Nakamura, Lisa, and Peter A. Chow-White, eds. *Race after the Internet*. New York: Routledge, 2011.

Namioka, Aki Helen, and Christopher Rao. "Introduction to Participatory Design." In *Field Methods Casebook for Software Design*, edited by Dennis Wixon and Judith Ramey, 283–299. New York: Wiley, 1996.

Nasaw, David. *Schooled to Order: A Social History of Public Schooling in the United States*. Oxford: Oxford University Press, 1979.

NationMaster. "Paraguay: Media Stats." Accessed November 6, 2018. http://www.nationmaster.com/country-info/profiles/Paraguay/Media.

Nava, Mura. "Some Obvious Notes on Mitra and Crawley (2014)." *EFL Notes* (blog), February 20, 2015. https://eflnotes.wordpress.com/2015/02/20/some-obvious-notes-on-mitra-and-crawley-2014/.

Negroponte, Nicholas. *Being Digital*. London: Hodder and Stoughton, 1996.

———. "Email Attachment," April 23, 2008. http://lists.laptop.org/pipermail/community-news/2008-April/000122.html, reposted on OLPC News at http://www.olpcnews.com/people/negroponte/nicholas_negroponte_sugar_olpc.html

———. "The Hundred Dollar Laptop: Computing for Developing Nations." Lecture at the Emerging Technologies Conference, MIT, Cambridge, MA, September 28, 2005. Video, 55:32, uploaded April 12, 2018. https://techtv.mit.edu/videos/16067-the-hundred-dollar-laptop-computing-for-developing-nations.

———. "The Hundred Dollar Man." Interview by Jason Pontin. *MIT Technology Review*, October 13, 2005. https://www.technologyreview.com/s/404814/the-hundred-dollar-man/.

———. "Laptop Project Viewed to Be More Cause than Opportunity." Interview by Dean Takahashi. *East Bay Times*, February 12, 2007. https://www.eastbaytimes.com/2007/02/12/laptop-project-viewed-to-be-more-cause-than-opportunity/.

———. "Laptops Work." Forum: Can Technology End Poverty? *Boston Review* 35, no. 6 (November/December 2010). http://bostonreview.net/archives/BR35.6/negroponte.php.

———. "Nicholas Negroponte and the $100 Laptop." Speech at Forrester Consumer Forum 2006, Chicago, IL, October 24–25, 2006. Audio, 38:06. Connected Social Media, November 3, 2006. https://connectedsocialmedia.com/2650/the-100-laptop/.

———. "Nicholas Negroponte at Digital, Life, Design." Transcript of remarks at the Digital, Life, Design Conference, Munich, Germany, January 21–23, 2007. OLPC Talks, 2007. https://web.archive.org/web/20070314085129/http://www.olpctalks.com:80/nicholas_negroponte/negroponte_digital_life_design.html

———. "Nicholas Negroponte at OAS - Presentation." Transcript of a lecture at the Organization of American States, Washington, DC, July 25, 2006. OLPC Talks, 2006. https://web.archive.org/web/20071024160254/http://www.olpctalks.com:80/nicholas_negroponte/negroponte_oas_presentation.html

———. "Nicholas Negroponte at OLPC Analyst Meeting (Interviews)." Transcript of remarks at analyst meeting, Cambridge, MA, April 26, 2007. OLPC Talks, 2007. https://web.archive.org/web/20071012165815/http://olpctalks.com:80/nicholas_negroponte/nicholas_negroponte_1.html

———. "Nicholas Negroponte at WCIT." Transcript of speech at the World Congress on Information Technology, Austin, TX, May 2006. OLPC Talks, 2006. https://web.archive.org/web/20071024160550/http://www.olpctalks.com:80/nicholas_negroponte/negroponte_wcit.html

———. "Nicholas Negroponte at WEF." Transcript of an interview by Loïc Le Meur at the World Economic Forum, Davos, Switzerland, January 2006. OLPC Talks, 2006. https://web.archive.org/web/20130411034406/http://www.olpctalks.com:80/nicholas_negroponte/negroponte_wef.html

———. "Nicholas Negroponte at World Bank Group." Transcript of a presentation at the World Bank Group, Washington DC, May 31, 2007. OLPC Talks, 2007. https://web.archive.org/web/20071014221537/http://www.olpctalks.com:80/nicholas_negroponte/negroponte_world_bank_group.html

———. "Nicholas Negroponte Keynote - Internet and Society 2007." Video recording of keynote address at Internet and Society 2007, Berkman Center for Internet and Society at Harvard Law School, Cambridge, MA, May 31, 2007. YouTube video, 50:41, published August 13, 2007. https://www.youtube.com/watch?v=or9GICoiXTo.

———. "No Lap Un-topped: The Bottom Up Revolution That Could Redefine Global IT Culture," Keynote Address, NetEvents Global Press Summit, Hong Kong, December 2, 2006. OLPC Talks, 2006. https://web.archive.org/web/20071012165720/http://olpctalks.com:80/nicholas_negroponte/negroponte_netevents.html

———. "The $100 Laptop." In *Globalization and Education*, edited by Marcelo Sánchez Sorondo, Edmond Malinvaud, and Pierre Léna, 19–23. Berlin: Walter de Gruyter, 2006.

———. "One Laptop per Child." Filmed February 2006 in Monterey, CA. TED video, 17:37. https://www.ted.com/talks/nicholas_negroponte_on_one_laptop_per_child.

———. "One-Room Rural Schools." *Wired*, September 1, 1998. https://www.wired.com/1998/09/negroponte-62/.

———. "Re-thinking Learning and Re-learning Thinking." Opening address at Learning Technologies Conference 2013, London, UK, January 29–30, 2013.

YouTube video, 57:43, published March 19, 2013. https://www.youtube.com/watch?v
=9K3Vmhjj2Gg.

———. "A 30-Year History of the Future." Filmed March 2014 in Vancouver, BC. TED
video, 19:40. https://www.ted.com/talks/nicholas_negroponte_a_30_year_history_of
_the_future.

Negroponte, Nicholas, and Walter Bender. "The New $100 Computer." World
Bank Group, May 31, 2007. OLPC Talks, 2007. https://web.archive.org/web/
20071014221537/http://www.olpctalks.com:80/nicholas_negroponte/negroponte
_world_bank_group.html.

Negroponte, Nicholas, and Chris Walsh. "Nicholas Negroponte at NECC 2006."
Transcript of a conversation at the National Educational Computing Conference, San Diego, CA, July 5–7, 2006. OLPC Talks, 2006. https://web.archive.org/
web/20071024160558/http://www.olpctalks.com:80/nicholas_negroponte/
nicholas_negroponte.html

Nestlé Uruguay. "Vascolet." Accessed November 6, 2018. http://www.nestle.com.uy/
productos/achocolatados/vascolet.

Nickson, Andrew. "The General Election in Paraguay, April 2008." *Electoral Studies*
28, no. 1 (March 2009): 145–149. https://doi.org/10.1016/j.electstud.2008.10.001.

Nissenbaum, Helen. "Hackers and the Contested Ontology of Cyberspace." *New Media
& Society* 6, no. 2 (April 2004): 195–217. https://doi.org/10.1177/1461444804041445.

Norman, Donald A., and Stephen W. Draper, eds. *User Centered System Design: New
Perspectives on Human-Computer Interaction*. 1986. Reprint, Boca Raton, FL: CRC,
2009.

Noss, Richard, and Celia Hoyles. *Windows on Mathematical Meanings: Learning Cultures and Computers*. Dordrecht, the Netherlands: Kluwer Academic, 1996.

Nye, David E. *American Technological Sublime*. Cambridge, MA: MIT Press, 1994.

Ogata, Amy F. *Designing the Creative Child: Playthings and Places in Midcentury America*. Minneapolis: University of Minnesota Press, 2013.

Oldenziel, Ruth. "Boys and Their Toys: The Fisher Body Craftsman's Guild, 1930–
1968, and the Making of a Male Technical Domain." *Technology and Culture* 38, no. 1
(January 1997): 60–96. http://doi.org/10.2307/3106784.

Olhero, Nelson, and Mark P. Sullivan. *Paraguay: Background and U.S. Relations*. CRS
Report for Congress, Order Code RL34180. Washington, DC: Congressional Research
Service, 2007.

Olivera, Oscar, in collaboration with Tom Lewis. *¡Cochabamba! Water War in Bolivia*.
Cambridge, MA: South End, 2004.

OLPC. "About the Laptop: Hardware." Accessed November 6, 2018. http://one.laptop
.org/about/hardware.

———. "Vision: History." Accessed November 6, 2018. http://laptop.org/en/vision/
project/index.shtml.

———. "Vision: Mission." Accessed November 6, 2018. http://laptop.org/en/vision/
mission/index.shtml.

OLPC News. "Negroponte: Criticising OLPC Is like Criticising the Church." September 15, 2006. http://www.olpcnews.com/people/negroponte/negroponte_to_critic.
html.

OLPC Wiki. "Activity Pack." Last edited January 18, 2009, 14:56. http://wiki.laptop.
org/go/Activity_pack.

———. "Deployments." Last edited June 22, 2013, 22:25. http://wiki.laptop.org/go/
Deployments.

———. "Disassembly." Last edited December 2, 2012, 23:44. http://wiki.laptop.org/
go/Disassembly.

———. "Game Jam." Last edited June 21, 2009, 19:42. http://wiki.laptop.org/go/
Game_Jam.

———. "Game Jam Boston June 2007." Last edited March 20, 2008, 04:21. http://
wiki.laptop.org/go/Game_Jam_Boston_June_2007.

———. "Games." Last edited August 7, 2013, 21:41. http://wiki.laptop.org/go/
Games.

———. "Hardware Uniqueness." Last edited May 17, 2011, 22:42. http://wiki.laptop
.org/go/Hardware_uniqueness.

———. "OLPC: Five Principles." Last edited February 6, 2013, 14:18. http://wiki
.laptop.org/go/OLPC:Five_principles.

———. "OLPC Myths." Last edited January 21, 2014, 03:20. http://wiki.laptop.org/
go/OLPC_myths.

———. "OLPC Principles and Basic Information." Last edited October 6, 2008,
03:56. http://wiki.laptop.org/go/OLPC_Principles_and_Basic_information.

———. "Our Market: Black Market." Last edited March 4, 1014, 14:51. http://wiki
.laptop.org/go/Our_market#Black_Market.

———. "Screws." Last edited July 18, 2018, 22:00. http://wiki.laptop.org/go/Screws.

———. "Talk: Activities." Last edited March 13, 2013, 19:13. http://wiki.laptop.org/
go/Talk: Activities.

———. "Video Chat." Last edited February 27, 2014, 22:01. http://wiki.laptop.org/ go/Video_Chat.

———. "XO Battery and Power." Last edited June 17, 2016, 21:30. http://wiki.laptop .org/go/Battery_and_power.

———. "XO-1 Touchpad: Issues." Last edited February 3, 2012, 04:50. http://wiki .laptop.org/go/XO-1/Touchpad/Issues.

Oppenheimer, Andres. "Latin America Leads in School Laptops." *Miami Herald,* March 21, 2011.

O'Reilly, Jessica. *The Technocratic Antarctic: An Ethnography of Scientific Expertise and Environmental Governance.* Ithaca, NY: Cornell University Press, 2017.

O'Shea, Tim, and John Self. *Learning and Teaching with Computers: Artificial Intelligence in Education.* Brighton, UK: Harvester, 1983.

O'Shea, Tim. "Mindstorms 2." Review of *The Children's Machine: Rethinking School in the Age of the Computer,* by Seymour Papert. *Journal of the Learning Sciences* 6, no. 4 (1997): 401–408. https://www.jstor.org/stable/1466779.

OSILAC (Observatory for the Information Society in Latin America and the Caribbean). *Characteristics of Households with ICTs in Latin America and the Caribbean.* United Nations Publication LC/W.171. Santiago, Chile: United Nations, 2007.

Oudshoorn, Nelly, Els Rommes, and Marcelle Stienstra. "Configuring the User as Everybody: Gender and Design Cultures in Information and Communication Technologies." *Science, Technology & Human Values* 29, no. 1 (January 2004): 30–63. https://doi.org/10.1177/0162243903259190.

Page, Angie S., Ashley R. Cooper, Pippa Griew, and Russell Jago. "Children's Screen Viewing Is Related to Psychological Difficulties Irrespective of Physical Activity." *Pediatrics* 126, no. 5 (November 2010): e1011–1017. https://doi.org/10.1542/ peds.2010-1154.

Palfrey, John, and Urs Gasser. *Born Digital: Understanding the First Generation of Digital Natives.* New York: Basic Books, 2008.

Papert, Seymour. *Mindstorms: Children, Computers, and Powerful Ideas.* New York: Basic Books, 1980.

———. *The Children's Machine: Rethinking School in the Age of the Computer.* New York: Basic Books, 1993.

———. "A Critique of Technocentrism in Thinking About the School of the Future." Seymour Papert (personal website). Accessed January 1, 2019. http://www.papert .org/articles/ACritiqueofTechnocentrism.html.

———. "Computer Criticism vs. Technocentric Thinking." *Educational Researcher* 16, no. 1 (January–February 1987): 22–30. https://doi.org/10.2307/1174251.

———. "Digital Development: How the $100 Laptop Could Change Education." Transcript of a USINFO Webchat, US Department of State, November 14, 2006. Library of Congress Web Archives Collection, archived November 13, 2008, 18:48. http://webarchive.loc.gov/all/20081113184800/http://usinfo.state.gov/usinfo/Archive/2006/Nov/14-358060.html.

Papert, Seymour, Clotilde Fonseca, Geraldine Kozberg, Michael Tempel, Sergei Soprunov, Elena Yakovleva, Horacio C. Reggini, Jeff Richardson, Maria Elizabeth B. Almeida, and David Cavallo. *Logo Philosophy and Implementation.* N.p.: Logo Computer Systems, 1999.

Papert, Seymour, and Idit Harel. "Situating Constructionism." In *Constructionism: Research Reports and Essays, 1985–1990*, edited by Seymour Papert and Idit Harel, 1–11. Norwood, NJ: Ablex, 1991.

Papert, Seymour, and Paulo Freire. "The Future of School." Seymour Papert (personal website). Accessed November 6, 2018. http://www.papert.org/articles/freire/freirePart1.html.

Pappas, Stephanie. "Exercise Doesn't Make Up for Kids' Screen Time." *Live Science*, October 11, 2010. https://www.livescience.com/8740-exercise-kids-screen-time.html.

Paraguay Educa Wiki. "Boletín Trimestral de Paraguay Educa, 2009 Verano." http://wiki.paraguayeduca.org/index.php/Boletin_2009_1 (site discontinued).

Parker, Edwin B. "Closing the Digital Divide in Rural America." *Telecommunications Policy* 24, no. 4 (May 2000): 281–290. https://doi.org/10.1016/S0308-5961(00)00018-5.

Patel, Nilay. "OLPC Hacked to Run Amiga OS." *Engadget*, January 10, 2008. https://www.engadget.com/2008/01/10/olpc-hacked-to-run-amiga-os/.

Patzer, Jeff. "Who's to Blame? Why the OLPC Plan in Peru Is Failing and Who Is Causing It." Pts. 1–6. Personal blog, January 1–6, 2011. https://web.archive.org/web/20110119070241/http://jeffpatzer.com:80/2011/01/

Paul, Annie Murphy. "The MOOC Gender Gap." *Slate*, September 10, 2014. https://slate.com/technology/2014/09/mooc-gender-gap-how-to-get-more-women-into-online-stem-classes.html.

Pawar, Udai Singh, Joyojeet Pal, Rahul Gupta, and Kentaro Toyama. "Multiple Mice for Retention Tasks in Disadvantaged Schools." In *Proceedings of the SIGCHI Conference on Human Factors in Computer System*, chaired by Mary Beth Rosson and David Gilmore, 1581–1590. New York: ACM, 2007. https://doi.org/10.1145/1240624.1240864.

Pea, Roy D. "The Aims of Software Criticism: Reply to Professor Papert." *Educational Researcher* 16, no. 5 (June 1987): 4–8. https://doi.org/10.3102/0013189X016005004.

———. "Cognitive Technologies for Mathematics Education." In *Cognitive Science and Mathematics Education*, edited by Alan H. Schoenfeld, 89–122. Hillsdale, NJ: Erlbaum, 1987.

———. "Symbol Systems and Thinking Skills: Logo in Context." In *Pre-proceedings of the 1984 National Logo Conference*, 55–61. Cambridge, MA: Laboratory for Computer Science, Massachusetts Institute of Technology, 1984.

Pea, Roy D., and D. Midian Kurland. *Logo Programming and the Development of Planning Skills*. Technical Report No. 16. New York: Bank Street College of Education, 1984. https://eric.ed.gov/?id=ED249930.

———. "On the Cognitive Effects of Learning Computer Programming." *New Ideas in Psychology* 2, no. 2 (1984): 137–168. https://doi.org/10.1016/0732-118X(84)90018-7.

Pea, Roy D., D. Midian Kurland, and Jan Hawkins. "LOGO and the Development of Thinking Skills." In *Children and Microcomputers: Research on the Newest Medium*, edited by Milton Chen and William Paisley, 193–212. Beverly Hills, CA: SAGE, 1985.

Pentagram. "One Laptop per Child: Brand Identity, Digital Design, Packaging." Accessed November 6, 2018. https://www.pentagram.com/work/one-laptop-per-child-1.

Phillips, Whitney. *This Is Why We Can't Have Nice Things: Mapping the Relationship between Online Trolling and Mainstream Culture*. Cambridge, MA: MIT Press, 2015.

Plan Ceibal. "Informe de evaluación del Plan Ceibal 2010." La Administración Nacional de Educación Pública (ANEP), March 2011. http://www.anep.edu.uy/anep-old/phocadownload/EvaluacionPlanCeibal/evaluacion%20del%20plan%20ceibal%20 2010%20resumen%20documento%20de%20iii%20%20marzo%202011.pdf

Poulsen, Kevin. "Negroponte: Laptop for Every Kid." *Wired*, November 17, 2005. https://www.wired.com/2005/11/negroponte-laptop-for-every-kid/.

Resnick, Mitchel. "Explore the Next Generation of Scratch." Workshop at Constructionism 2012: Theory Practice and Impact, Metropolitan Hotel, Athens, Greece, August 22, 2012.

———. *Lifelong Kindergarten: Cultivating Creativity through Projects, Passion, Peers, and Play*. Cambridge, MA: MIT Press, 2017.

Resnick, Mitchel, John Maloney, Andrés Monroy-Hernández, Natalie Rusk, Evelyn Eastmond, Karen Brennan, Amon Millner, et al. "Scratch: Programming for All." *Communications of the ACM* 52, no. 11 (November 2009): 60–67. https://doi.org/10.1145/1592761.1592779.

Rheingold, Howard. "A Slice of Life in My Virtual Community." In *Global Networks: Computers and International Communication*, edited by Linda M. Harasim, 57–80. Cambridge, MA: MIT Press, 1993.

Richtel, Matt. "Wasting Time Is New Divide in Digital Era." *New York Times*, May 29, 2012. https://www.nytimes.com/2012/05/30/us/new-digital-divide-seen-in-wasting -time-online.html.

Rousseau, Jean Jacques. *Émile; or, Concerning Education: Extracts Containing the Principal Elements of Pedagogy Found in the First Three Books*. Translated by Eleanor Worthingon. Boston: D. C. Heath, 1889.

Rury, John L. *Education and Social Change: Themes in the History of American Schooling*. 2nd ed. Mahwah, NJ: Lawrence Erlbaum Associates, 2005.

Salamano, Ignacio, Pablo Pagés, Analí Baraibar, Helena Ferro, Laura Pérez, and Martín Pérez. "Monitoreo y evaluación educativa del Plan Ceibal: Primeros resultados a nivel nacional." Plan Ceibal, December 2009. https://www.ceibal.edu.uy/ storage/app/uploads/public/58a/4ab/2e1/58a4ab2e1b524319302667.pdf.

Santiago, Ana, Eugenio Severin, Julian Cristia, Pablo Ibarrarán, Jennelle Thompson, and Santiago Cueto. *Experimental Assessment of the Program "One Laptop per Child" in Peru*. IDB Education Briefly Noted No. 5. Washington, DC: Inter-American Development Bank, 2010. https://publications.iadb.org/handle/11319/3876.

Scheper-Hughes, Nancy. *Death without Weeping: The Violence of Everyday Life in Brazil*. Berkeley: University of California Press, 1993.

Schofield, Jack. "Late News: Seymour Papert Injured in Vietnam." *Guardian*, January 3, 2007. https://www.theguardian.com/technology/blog/2007/jan/03/ latenewsseymo.

Schuler, Douglas, and Aki Namioka, eds. *Participatory Design: Principles and Practices*. Hillsdale, NJ: Lawrence Erlbaum Associates, 1993.

SciDev.Net. "Paraguay ampliaría plan de computadoras para niños." May 8, 2011. https://www.scidev.net/america-latina/brecha-digital/noticias/paraguay-ampliar-a -plan-de-computadoras-para-ni-os.html.

Segal, Howard P. *Technological Utopianism in American Culture*. Twentieth anniversary edition. Syracuse, NY: Syracuse University Press, 2005.

Self, John. *Microcomputers in Education: A Critical Evaluation of Educational Software*. Brighton, UK: Harvester, 1985.

Selwyn, Neil. "The Digital Native: Myth and Reality." *Aslib Proceedings* 61, no. 4 (2009): 364–379. https://doi.org/10.1108/00012530910973776.

Shane, Scott. "Cables Show Central Negroponte Role in 80's Covert War against Nicaragua." *New York Times*, April 13, 2005. https://www.nytimes.com/2005/04/13/politics/cables-show-central-negroponte-role-in-80s-covert-war-against.html.

Shreeve, Jimmy Lee. "Hand-Cranked Computers: Is This a Wind-Up?" *Independent*, November 23, 2005. https://www.independent.co.uk/news/science/hand-cranked -computers-is-this-a-wind-up-516594.html.

Sims, Christo. *Disruptive Fixation: School Reform and the Pitfalls of Techno-idealism*. Princeton, NJ: Princeton University Press, 2017.

———. "Video Game Culture, Contentious Masculinities, and Reproducing Racialized Social Class Divisions in Middle School." *Signs: Journal of Women in Culture and Society* 39, no. 4 (Summer 2014): 848–857. https://doi.org/10.1086/675539.

Smith, Sylvia. "The $100 Laptop: Is It a Wind-Up?" *CNN*, December 1, 2005. http://edition.cnn.com/2005/WORLD/africa/12/01/laptop/.

Srinivasan, Janaki, and Jenna Burrell. "Revisiting the Fishers of Kerala, India." In *Proceedings of the Sixth International Conference on Information and Communication Technologies and Development*, chaired by Gary Marsden, Julian May, Jonathan Donner, and Tapan Parikh, 56–66. New York: ACM, 2013. https://doi .org/10.1145/2516604.2516618.

Stack Overflow. "2015 Developer Survey." Accessed November 6, 2018. http://stackoverflow.com/research/developer-survey-2015.

Stahl, Lesley. "What if Every Child Had a Laptop." Produced by Catherine Olian. *60 Minutes*. Aired May 20, 2007, on CBS. Video, 12:39. https://www.cbsnews.com/news/what-if-every-child-had-a-laptop/.

Stahl, William A. *God and the Chip: Religion and the Culture of Technology*. Waterloo, ON: Wilfrid Laurier University Press for the Canadian Corporation for Studies in Religion, 1999.

Star, Susan Leigh. "The Ethnography of Infrastructure." *American Behavioral Scientist* 43, no. 3 (November 1999): 377–391. https://doi.org/10.1177/00027649921955326.

———. "This Is Not a Boundary Object: Reflections on the Origin of a Concept." *Science, Technology & Human Values* 35, no. 5 (September 2010): 601–617. https://doi.org/10.1177/0162243910377624.

Stavrinaki, Maria. "The African Chair or the Charismatic Object." *Grey Room*, no. 41 (Fall 2010): 88–110. https://doi.org/10.1162/GREY_a_00011.

Suchman, Lucy A. *Plans and Situated Actions: The Problem of Human-Machine Communication*. Cambridge: Cambridge University Press, 1987.

Sullivan, Nick. "The Man behind Logo." *Family Computing* 2, no. 2 (February 1984): 70–71.

Sunstein, Cass R. *#Republic: Divided Democracy in the Age of Social Media.* Princeton, NJ: Princeton University Press, 2017.

———. *Republic.com.* Princeton, NJ: Princeton University Press, 2002.

———. *Republic.com 2.0.* Princeton, NJ: Princeton University Press, 2007.

Takhteyev, Yuri. *Coding Places: Software Practice in a South American City.* Cambridge, MA: MIT Press, 2012.

———. "Networks of Practice as Heterogeneous Actor-Networks: The Case of Software Development in Brazil." *Information, Communication & Society* 12, no. 4 (2009): 566–583. https://doi.org/10.1080/13691180902859369.

Talbot, David. "Given Tablets but No Teachers, Ethiopian Children Teach Themselves." *MIT Technology Review*, October 29, 2012. https://www.technologyreview .com/s/506466/given-tablets-but-no-teachers-ethiopian-children-teach-themselves/.

Tate, Paul. "The Blossoming of European AI." *Datamation*, International Edition, November 1, 1984, 85–88.

Taylor, Charles. *Modern Social Imaginaries.* Durham, NC: Duke University Press, 2004.

Thomas, Douglas. *Hacker Culture.* Minneapolis: University of Minnesota Press, 2002.

Thompson, Iain. "African Kids Learn to Read, Hack Android on OLPC Fondleslab." *Register*, November 1, 2012. https://www.theregister.co.uk/2012/11/01/kids_learn _hacking_android/.

———. "Negroponte Plans Tablet Airdrops to Teach Kids to Read." *Register*, November 2, 2011. https://www.theregister.co.uk/2011/11/02/negroponte_tablet_airdrops/.

Thornburg, David D. "Whatever Happened to Logo?" *Compute!*, no. 78 (November 1986): 83.

TNW. "Why Half of Developers Don't Have a Computer Science Degree." April 23, 2016. https://thenextweb.com/insider/2016/04/23/dont-need-go-college-anymore -programmer/.

Torn Halves. "Sugata Mitra on EdTech and Empire." *Digital Counter-Revolution*, March 8, 2013. http://www.digitalcounterrevolution.co.uk/2013/sugata-mitra-edtech-empire -ted-prize-talk.

Toyama, Kentaro. "Can Technology End Poverty?" *Boston Review* 35, no. 6 (November/December 2010). http://bostonreview.net/archives/BR35.6/toyama.php.

———. *Geek Heresy: Rescuing Social Change from the Cult of Technology.* New York: PublicAffairs, 2015.

Transparency International. "Paraguay." Accessed November 6, 2018. https://www
.transparency.org/country/PRY.

Tsing, Anna. "The Global Situation." *Cultural Anthropology* 15, no. 3 (August 2000):
327–360. https://doi.org/10.1525/can.2000.15.3.327.

Turkle, Sherry, and Seymour Papert. "Epistemological Pluralism: Styles and Voices
within the Computer Culture." *Signs: Journal of Women in Culture and Society* 16,
no. 1 (Autumn 1990): 128–157. https://doi.org/10.1086/494648.

Turner, Fred. *The Democratic Surround: Multimedia and American Liberalism from
World War II to the Psychedelic Sixties*. Chicago: University of Chicago Press, 2013.

———. *From Counterculture to Cyberculture: Stewart Brand, the Whole Earth Network,
and the Rise of Digital Utopianism*. Chicago: University of Chicago Press, 2006.

———. "Where the Counterculture Met the New Economy: The WELL and the
Origins of Virtual Community." *Technology and Culture* 46, no. 3 (2005): 485–512.
https://doi.org/10.1353/tech.2005.0154.

———. "How Digital Technology Found Utopian Ideology: Lessons from the First
Hackers' Conference." In *Critical Cyberculture Studies*, edited by David Silver and
Adrienne Massanari, 257–259. New York: New York University Press, 2006.

———. "Burning Man at Google: A Cultural Infrastructure for New Media Production." *New Media & Society* 11, no. 1–2 (February 2009): 73–94. https://doi.org/
10.1177/1461444808099575.

Twist, Jo. "UN Debut for $100 Laptop for Poor." *BBC News*, November 17, 2005.
http://news.bbc.co.uk/2/hi/technology/4445060.stm.

Tyack, David, and Larry Cuban. *Tinkering toward Utopia: A Century of Public School
Reform*. Cambridge, MA: Harvard University Press, 1995.

UN News. "UN Supports Project Aimed at Providing Cheap Laptops to Students in
Poor Countries." January 28, 2006. https://news.un.org/en/story/2006/01/167492
-un-supports-project-aimed-providing-cheap-laptops-students-poor-countries.

UNESCO Institute for Statistics. "Literacy Rate, Adult Total (% of People Ages 15
and Above)." World Bank (website). Accessed November 6, 2018. https://data
.worldbank.org/indicator/SE.ADT.LITR.ZS.

Vessuri, Hebe M. C. "The Social Study of Science in Latin America." *Social
Studies of Science* 17, no. 3 (August 1987): 519–554. https://doi.org/10.1177
%2F030631287017003006.

Vigdor, Jacob L., Helen F. Ladd, and Erika Martinez. "Scaling the Digital Divide:
Home Computer Technology and Student Achievement." *Economic Inquiry* 52, no. 3
(July 2014): 1103–1119. https://doi.org/10.1111/ecin.12089.

Vota, Wayan. "How to ... Use Failures to Succeed in Technology for Development." *Guardian*, December 7, 2012. https://www.theguardian.com/global-development -professionals-network/2012/dec/07/fail-faire-how-to-talk-about-failure.

———. "Is OLPC the Only Hope to Eliminate Poverty and Create Peace?" OLPC News, November 30, 2007. http://www.olpcnews.com/people/negroponte/olpc _poverty_world_peace.html.

———. "Negroponte Throws XO on Stage Floor 'Do This with a Dell'—Everyone a Twitter." OLPC News Forums, 2009. Accessed May 1, 2013. http://olpcnews.com/ forum/?topic=4597.0 (forum discontinued).

———. "OLPC Mission Change: Constructionism & Competition, Gone!" OLPC News, April 11, 2007. http://www.olpcnews.com/commentary/olpc_news/olpc _mission_constructionism.html.

———. "One Laptop per Child 'Trojan Horse' Comparison." OLPC News, May 11, 2007. http://www.olpcnews.com/people/negroponte/laptop_child_trojan_horse .html.

———. "One Pornographic Image per Nigerian Child." OLPC News, July 19, 2007. http://www.olpcnews.com/countries/nigeria/pornographic_image_child.html.

———. "XO Helicopter Deployments? Nicholas Negroponte Must Be Crazy!" OLPC News, June 29, 2011. http://www.olpcnews.com/people/negroponte/xo _helicopter_deployments_nich.html.

W3Techs. "Usage of Content Languages for Websites." Accessed November 6, 2018. https://w3techs.com/technologies/overview/content_language/all.

WageIndicator Foundation. "Salarios mínimos en Paraguay desde el 01-07-2011 al 28-02-2014." Tu Salario (website), June 30, 2014. https://tusalario.org/paraguay/ salario/salario-minimo-1/archive/0/.

Walter-Herrmann, Julia, and Corinne Büching, eds. *FabLab: Of Machines, Makers and Inventors*. Bielefeld, Germany: Transcript-Verlag, 2014.

Wang, Ping, Han-Hui Liu, Xing-Ting Zhu, Tian Meng, Hui-Jie Li, and Xi-Nian Zuo. "Action Video Game Training for Healthy Adults: A Meta-analytic Study." *Frontiers in Psychology* 7 (June 17, 2016). https://doi.org/10.3389/fpsyg.2016.00907.

Ward, Janelle. "A Critique of Hole-in-the-Wall (HiWEL)." *Broker*, July 3, 2010. http:// www.thebrokeronline.eu/Blogs/Janelle-Ward/A-critique-of-Hole-in-the-Wall-HiWEL.

Warren, Tom. "Chromebooks Outsold Macs for the First Time in the US." *Verge*, May 19, 2016. https://www.theverge.com/2016/5/19/11711714/chromebooks-outsold-macs -us-idc-figures.

Warschauer, Mark. "Demystifying the Digital Divide." *Scientific American* 289, no. 2 (August 2003): 42–47. https://doi.org/10.1038/scientificamerican0803-42.

Warschauer, Mark, and Morgan G. Ames. "Can One Laptop per Child Save the World's Poor?" *Journal of International Affairs* 64, no. 1 (Fall/Winter 2010): 33–51. https://www.jstor.org/stable/24385184.

Warschauer, Mark, and T. Matuchniak. "New Technology and Digital Worlds: Analyzing Evidence of Equity in Access, Use, and Outcomes." *Review of Research in Education* 34 (2010): 179–225. https://doi.org/10.3102/0091732X09349791.

Watters, Audrey. "Click Here to Save Education: Evgeny Morozov and Ed-Tech Solutionism." *Hack Education* (blog), March 26, 2013. http://hackeducation.com/2013/03/26/ed-tech-solutionism-morozov.

Webb, N. M., P. Ender, and S. Lewis. "Problem-Solving Strategies and Group Processes in Small Groups Learning Computer Programming." *American Educational Research Journal* 23, no. 2 (June 1986): 243–261. https://doi.org/10.3102/00028312023002243.

Weber, Max. "Charismatic Authority." In *The Theory of Social and Economic Organization*, edited by Talcott Parsons, 358–363. Translated by A. M. Henderson and Talcott Parsons. Glencoe, IL: Free Press, 1947.

Westervelt, Eric. "The Cubist Revolution: Minecraft For All." *Morning Edition*. Aired August 8, 2017, on NPR. Audio, 3:52. https://www.npr.org/sections/ed/2017/08/08/538580856/the-cubist-revolution-minecraft-for-all.

Wilby, Peter. "Sugata Mitra: The Professor with His Head in the Cloud." *Guardian*, June 7, 2016. https://www.theguardian.com/education/2016/jun/07/sugata-mitra-professor-school-in-cloud.

Willis, Paul. *Learning to Labor: How Working-Class Kids Get Working-Class Jobs*. 1977. Reprint, New York: Columbia University Press, 1981.

Wilson, Kenneth R., Jennifer S. Wallin, and Christa Reiser. "Social Stratification and the Digital Divide." *Social Science Computer Review* 21, no. 2 (May 2003): 133–143. https://doi.org/10.1177%2F0894439303021002001.

Winner, Langdon. "Do Artifacts Have Politics?" In *The Whale and the Reactor: A Search for Limits in an Age of High Technology*, 19–39. Chicago: University of Chicago Press, 1986.

———. "Technology Today: Utopia or Dystopia?" *Social Research* 64, no. 3 (Fall 1997): 989–1017. https://www.jstor.org/stable/40971195.

———. "Upon Opening the Black Box and Finding It Empty: Social Constructivism and the Philosophy of Technology." *Science, Technology & Human Values* 18, no. 3 (Summer 1993): 362–378. https://doi.org/10.1177/016224399301800306.

Winston, Brian. "Let Them Eat Laptops: The Limits of Technicism." *International Journal of Communication* 1 (2007): 170–176. https://ijoc.org/index.php/ijoc/article/view/150/72.

Wolf, Eric R. *Europe and the People without History.* Berkeley: University of California Press, 1982.

World Bank. *Paraguay: Estudio de pobreza: Determinantes y desafíos para la reducción de la pobreza.* Report No. 58638-PY. Washinton, DC: World Bank, 2010. http://documents.worldbank.org/curated/en/216291468145457392/Paraguay-Estudio-de-pobreza-determinantes-y-desafios-para-la-reduccion-de-la-pobreza.

Ybema, Sierk, and Frans Kamsteeg. "Making the Familiar Strange: A Case for Disengaged Organizational Ethnography." In Ybema, Sierk, Dvora Yanow, Harry Wels, and Frans Kamsteeg (eds), *Organizational Ethnography: Studying the Complexities of Everyday Life*, 101–119. London: SAGE Publications, 2009. https://doi.org/10.4135/9781446278925.

Yelland, Nicola. "Mindstorms or a Storm in a Teacup? A Review of Research with Logo." *International Journal of Mathematical Education in Science and Technology* 26, no. 6 (1995): 853–869. https://doi.org/10.1080/0020739950260607.

Zornado, Joseph L. *Inventing the Child: Culture, Ideology, and the Story of Childhood.* New York: Routledge, 2006.

Zuckerman, Ethan. "Child's Play: How One Laptop per Child Plans to Bring Computers to a Billion Schoolchildren ... and a Revolution to the Computer Industry." *... My Heart's in Accra* (blog), January 19, 2007. http://www.ethanzuckerman.com/blog/childs-play-how-one-laptop-per-child-plans-to-bring-computers-to-a-billion-schoolchildren-and-a-revolution-to-the-computer-industry/.

———. "It's Cute. It's Orange. It's Got Bunny Ears. An Update on the One Laptop per Child Project." *... My heart's in Accra* (blog), June 1, 2006. http://www.ethanzuckerman.com/blog/2006/06/01/its-cute-its-orange-its-got-bunny-ears-an-update-on-the-one-laptop-per-child-project/

Index

Note: Figures are indicated by "f" following page numbers.